INVESTING IN THE ERA OF CLIMATE CHANGE

INVESTING IN THE ERA OF CLIMATE CHANGE

BRUCE USHER

Columbia Business School
Publishing

Columbia University Press
Publishers Since 1893
New York Chichester, West Sussex
cup.columbia.edu

Library of Congress Cataloging-in-Publication Data
Names: Usher, Bruce, author.
Title: Investing in the era of climate change / Bruce Usher.
Description: New York, NY : Columbia University Press, [2022] |
Includes bibliographical references and index.
Identifiers: LCCN 2022004785 (print) | LCCN 2022004786 (ebook) |
ISBN 9780231200882 (hardback) | ISBN 9780231553827 (ebook)
Subjects: LCSH: Investments—Environmental aspects. |
Climatic changes—Economic aspects.
Classification: LCC HG4521 .U84 2022 (print) | LCC HG4521 (ebook) |
DDC 332.6—dc23/eng/20220321
LC record available at https://lccn.loc.gov/2022004785
LC ebook record available at https://lccn.loc.gov/2022004786

Cover design: Noah Arlow

CONTENTS

PREFACE

Climate change is a controversial topic. Just over one in two Americans believe global warming is caused mostly by human activity, and there is even less agreement on what should be done about it.[1] In fact, the science is unambiguous: human emissions of greenhouse gases have warmed the Earth by 1°C in the past century, and the current emissions trajectory will result in 4°C–6°C of warming by the end of this century, with potentially disastrous consequences for humanity.[2]

The science is similarly unambiguous on what must be done to avoid catastrophe: reduce emissions of greenhouse gases to zero, ideally by 2050 and no later than 2070. Fortunately, there is a path for achieving this. Commercial solutions already exist to reduce emissions by more than half, and technologies are under development to tackle the rest.[3] But avoiding catastrophe will require implementation of climate solutions at an unprecedented pace and at a global scale, and that will require an extraordinary amount of investment capital: an estimated $125 trillion by 2050 to decarbonize the global economy and limit warming to an adaptable 1.5°C–2°C.[4]

Investing in the Era of Climate Change focuses on the implications for investors and the role that investment plays in addressing climate change. Governments also have a critical role, ensuring emissions are regulated and capital is properly incentivized, yet engagement of the private sector is critical to implement climate solutions at the scale and speed needed to address the climate challenge.

Climate change is global, and the implications for investors are global as well. However, to provide focus, this book is written from an American perspective. Investors in other countries will experience similar trends, challenges, and opportunities as in the United States, and the lessons are applicable globally. Note that *Investing in the Era of Climate Change* does not explain every climate solution or technology, nor every investment product or opportunity, as there are too many to include in a single volume.

I began investing in climate change solutions in 2002. At that time, there were few investment opportunities because climate solutions were costly and uncompetitive without generous government subsidies. Twenty years later, much has changed. Climate change is significantly worse, and less time remains to address it. But the technologies and business models for reducing emissions have dramatically improved, offering a path for avoiding catastrophic climate change. *Investing in the Era of Climate Change* describes that path and the critical role that investors play in navigating it. My hope in writing this book is that every investor—individual and institutional—will recognize the changes that are coming and act, for their benefit and the benefit of all.

INVESTING IN THE ERA OF CLIMATE CHANGE

SECTION 1

Momentum

Momentum: strength or force gained by motion or by a series of events.

MERRIAM-WEBSTER DICTIONARY[1]

FIGURE 1.1. The Watt Steam Engine (*Source*: Old Book Illustrations)

Chapter One

PROSPERITY, WITH A CATCH

James Watt invented the steam engine in 1723, creating a turning point in the history of humanity. For the first time, people could harness mechanical energy from fossil fuels, raising productivity beyond the limitations of brute human or animal endurance. Watt's invention helped power the Industrial Revolution by replacing human labor in manufacturing and animals in transportation, accelerating a burst of technological innovation that continues to this day. Factories, utilities, railways, and ships were designed and built using the technologies invented by Watt and many others that followed. This required ingenuity, perseverance, and luck. It also required vast amounts of capital.

The Industrial Revolution created an unprecedented demand for investment to build the machinery that produced the world's goods and transported them to market. Capital became a key driver of economic growth and prosperity, first in Britain and later in Continental Europe and America. Banks, which provided much of the capital required for the Industrial Revolution, became "the pillar of Britain's industrial edifice."[1] In the United States, the Industrial Revolution required massive financing for infrastructure development, especially for the canals and railways that supported trade across the immense new country.

The Industrial Revolution was followed by the Agricultural Revolution, a lesser known but equally important development in human prosperity. Advances in farming practices and agricultural machinery increased the productivity of farmers throughout Europe and the United States, reducing the risk of starvation and malnutrition. In 1910, two German chemists invented the Haber-Bosch process, which led to the development of synthetic nitrogen fertilizer. Haber won the Nobel Prize in Chemistry for his work, reflecting the importance of his discovery to humanity.[2] The use of nitrogen fertilizer created a global revolution in agriculture—in the United States, corn yields per acre grew 500 percent, allowing farmers to grow five times as much food on the same plot of land.[3] Globally, synthetic nitrogen fertilizers doubled agricultural production in a few short decades, a remarkable achievement after twelve millennia of slow gains.[4] The human population, no longer limited by food shortages, expanded rapidly, from approximately 700 million to more than 7 billion.[5]

With the Industrial and Agricultural Revolutions came a sustained increase in productivity and incomes, the first in recorded history. Prior to the eighteenth century, everyone other than a small ruling class lived in abject poverty. Real gross domestic product (GDP) per capita in Europe was slightly more than $3 per day from earliest recorded history through 1200 CE, slowly rising to a mere $6 per day by the time James Watt invented the steam engine.[6] The Industrial and Agricultural Revolutions led to a dramatic increase in incomes, and not just for the wealthy. By the twenty-first century, real GDP per capita in Europe and the United States was nearly 50 times higher than in pre-industrial times.[7]

For the first time in history, human prosperity materially improved, thanks to better health, less food insecurity, and an abundance of goods and services that would have once seemed unimaginable. But there was a catch.

CLIMATE CHANGE

The Industrial Revolution was powered by fossil fuels—burning vast quantities of coal, emitting carbon dioxide into the atmosphere, and warming the planet. Watt's steam engine ran on coal, as did much of the machinery and transportation that underpinned early economic growth. In the twentieth century, oil grew in importance to power automobiles, airplanes, and ships. The Agricultural Revolution also contributed to climate change, as nitrogen fertilizers emit a potent greenhouse gas. The expansion of agricultural land to serve a growing population drove massive deforestation, releasing carbon previously stored in trees, while the livestock that graze these deforested lands release methane through their digestion.

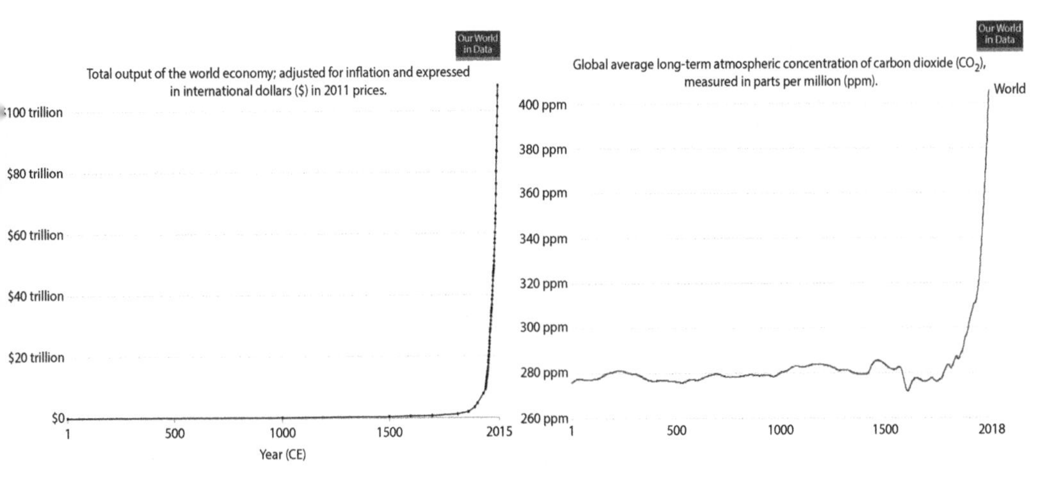

FIGURE 1.2. (*left*) Growth of world GDP over the past two millennia. (*right*) Increase in CO_2 concentration over the past two millennia. (Source: Our World in Data)

The Industrial and Agricultural Revolutions created unprecedented economic growth that has dramatically raised incomes and reduced poverty and hunger for most of the world's peoples. But in doing so, greenhouse gases were emitted into the atmosphere in record volume creating anthropogenic (human-caused) climate change. The connection between GDP growth and atmospheric CO_2 concentration is not coincidental, it is causal (figure 1.2).

Scientists have long known of the link between economic growth and emissions and have spent decades warning governments of the rising risk of catastrophic climate change. The most recent consensus from 14,000 climate scientists forecasts average warming as high as 6°C by the end of the century.[8] In some regions, temperatures are forecast to rise much more, with the Arctic expected to warm by up to three times the average.[9] By 2050, within the lifetimes of most people, the planet will be warming faster than it has for tens of millions of years.[10]

WHAT MUST BE DONE

Climate scientists are clear on what must be done if humanity is to avoid a catastrophe: reduce global greenhouse gas emissions to zero, ideally by 2050 to limit warming to 1.5°C and no later than 2070 to keep temperature rise within 2°C.[11] Not meeting the science-based targets for net zero emissions risks dramatic changes to the earth's natural environment that will reverse the prosperity gains of the Industrial and Agricultural Revolutions and leave humanity struggling on a planet with rising seas, extreme weather events, and punishing heat waves. Even a small gain in temperature will result in a large change in outcomes.

NASA's James Hansen was the first scientist to warn government leaders of climate change risk, speaking before Congress in 1988.[12] In the decades since then, scientists have prepared increasingly detailed and accurate forecasts of climate change for government officials, providing a clear mitigation path for policy makers to follow. Unfortunately, political leaders have proven unable to effectively act on those warnings, failing to halt the rise in global greenhouse gas emissions. Much of that failure has been blamed on politics. In truth, climate change poses a uniquely challenging problem for government to solve.

THE ROLE OF GOVERNMENT

Climate change is a *negative externality*, a term used by economists to describe how the production or consumption of a good has a harmful side effect on others. For instance, the burning of coal to generate electricity emits CO_2 into the atmosphere, contributing to the warming of the planet, but neither the company generating the electricity nor the consumer using it must bear the burden of their actions. In the case of climate change, the negative externality is experienced by subsequent generations, as there is a lag between the emission of greenhouse gases and the warming of the planet.

Negative externalities are addressed through government intervention, in the form of regulations to limit or prohibit the polluting activity or with a tax that forces the polluter to pay for the harm done. A classic example is leaded gasoline, which was once common in America. Adding lead to gasoline enhanced engine performance in automobiles, improving the experience for drivers. Lead was known to be poisonous, but its impact on health took years to appear. A public health report written in 1926 concluded that exposure to lead poisoning "would be another generation's problem."[13] And so it was. By the 1970s, lead poisoning from automobile emissions was damaging the health of millions of American children, including retardation in growth and development. Under pressure from health authorities, the U.S. government enacted regulations to reduce and eventually eliminate leaded gasoline, thereby addressing the negative externality.[14]

But leaded gasoline was an easier problem to solve than climate change because the negative externality was local—the automobiles emitting lead were driven in the same communities as the children harmed—and thus a U.S. government regulation could solve the problem. Climate change is different. Emissions of CO_2 and other greenhouse gases rise into the earth's atmosphere where the molecules disperse over the entire planet, amplifying the greenhouse effect. Emissions of greenhouse gases from *anywhere* on the planet affect *everyone* on the planet. Solving a negative externality such as climate change requires an understanding of the economics of shared resources.

THE TRAGEDY OF THE COMMONS

Two centuries ago, British economist William Forster Lloyd noticed that when farmers allowed their cows to graze on public (or common) land, the farmer had an incentive to allot as many animals as possible. He surmised that if each and every farmer did what was in his individual best interest, the public land would be damaged through overuse, and every farmer would suffer. Lloyd concluded that rational economic decisions made by individuals will eventually deplete or destroy common resources to the detriment of the group. Even though the individuals know this will occur, it remains in their best interest to take advantage of the common space. This concept became known as the *tragedy of the commons*, and it applies to climate change. The farmers are every business and individual on the planet that emits greenhouse gases, and the commons is the earth's atmosphere. It is a very large commons, but not infinitely large.

Governments address the tragedy of the commons problem through property rights. In the example of grazing, the government can simply limit the number of cows each farmer is allowed to put onto the commons or it can charge a tax per cow equivalent to the damage it causes on the commons and use the tax revenue to repair any damage done. In either case, the farmer's individual incentives will align with the needs of the community, and the commons will be protected. In the case of climate change, governments simply need to regulate or tax emissions of greenhouse gases to change the behavior of individuals and businesses responsible for the pollution. But the regulation or tax must apply to *everyone* using the commons. With climate change, everyone using the commons is literally everyone on the planet. Which means *national* governments must work together to reach an effective *international* agreement, and that is, to put it mildly, difficult.

ATTEMPTS TO ADDRESS CLIMATE CHANGE WITH INTERNATIONAL AGREEMENTS

In 1992, governments representing 172 countries met in Rio de Janeiro to negotiate the first international treaty to address climate change. The resulting agreement, the United Nations Framework Convention on Climate Change, created a structure for subsequent negotiations. Beginning in 1995, government negotiators met annually to hammer out the terms for international regulation of greenhouse emissions. The 1997 meeting in Japan resulted in the first binding agreement, called the Kyoto Protocol, which capped emissions in developed countries. Unfortunately, the countries that agreed to a cap subsequently backtracked, and the agreement expired in 2012 after producing little effect on climate change. After several unsuccessful attempts, in 2015 negotiators settled on a new deal, the Paris Agreement, with much fanfare.

The Paris Agreement to address climate change was signed by 195 countries. Nations responsible for more than 90 percent of global emissions agreed to submit carbon-reduction targets outlining each country's commitment to curb emissions. The Paris Agreement was lauded for finally bringing nearly every country on the planet into one concerted plan to tackle the great tragedy of the commons that is climate change. President Obama declared "history may well judge it as a turning point for our planet."[15]

But the Paris Agreement neither caps emissions nor penalizes countries for missing targets. There are no specific requirements as to how much each country must reduce emissions, and the goals set by each country are not legally binding. In the years since the Paris Agreement was signed, global emissions of greenhouse gases have continued to increase, rising from nearly 53 gigatonnes of CO_2 equivalent (Gt CO_2e) in 2014[16] before the agreement was signed to more than 59 Gt CO_2e in 2019.[17] Emissions declined for the first time in 2020, but only due to the global COVID-19 pandemic, which restricted travel and economic activity. The United Nations noted "this economic disruption has briefly slowed—but far from eliminated—the historic and ever-increasing burden of human activity on the Earth's climate."[18]

INVESTOR REACTION

Investors have looked to international agreements for direction on where and how to allocate capital to mitigate and adapt to climate change. This makes sense, as the fundamental challenge of climate change—the tragedy of the commons—has historically been addressed with government regulations to direct investment capital. In the case of climate change, however, the absence of policies with clear and binding rules has made it nearly impossible for businesses and investors to make long-term capital commitments. The rise and fall of politicians have further contributed to the lack of stability; American presidents have twice entered international climate agreements only to see subsequent administrations unilaterally pull out.

Governments have attempted to address climate change for more than 30 years through negotiated international agreements, including landmark agreements in 1997 and 2015. It has not worked. Solving the tragedy of the commons that is climate change has been stymied by the need for a global consensus among nations, an unlikely outcome in a politically fractious world. The absence of stable climate policy convinced most business leaders and investors to take the most cautious course of action, which was to do nothing. Fortunately, that is now changing and changing fast.

A TURNING POINT

In 2021, business leaders switched tack, stepping up to announce plans to address climate change. More than 3,000 companies, including many of the world's largest, committed to net zero greenhouse gas emissions, pledging a rapid and sustained decrease in line with science-based targets. Investors responded with enthusiastic support.

At the annual meeting of climate negotiators in 2021, members of the financial community announced the Glasgow Financial Alliance for Net Zero, a global coalition of 450 financial firms managing assets of more than $130 trillion that are committed to reducing greenhouse gas emissions to zero.[19] This commitment explicitly builds on the 2015 Paris Agreement, demonstrating how international agreements can provide a framework for businesses to build upon, even in the absence of binding policies. The *Wall Street Journal* summed up the tenor of the meeting with the headline "Business Is the Game-Changer at COP26 in Glasgow."[20] Business leaders and investors are, at long last, joining government officials in plans to reduce emissions and tackle climate change.

This new direction by the private sector, working with and in some cases ahead of government, was met with surprise and skepticism by many observers. In fact, there are very good reasons for this turning point. Just as the Industrial Revolution spurred capital formation in industry, several global trends are driving a rapid and sustained flow of capital into climate solutions. Investing in the era of climate change has begun.

"I want you to act as if our house is on fire, because it is."

FIGURE 2.1. Greta Thunberg, climate activist (*Source*: Wikimedia Commons)

Chapter Two

INVESTING IN THE ERA OF CLIMATE CHANGE

Climate scientists have been sounding the alarm about the need for investment to reduce greenhouse gas emissions, but investors have been slow to commit capital because mitigation of climate change requires collective action. Fortunately for the planet, recent trends are prompting financial leaders to finally act. Understanding those trends and the implications for investors is the first step toward investing in the era of climate change.

TREND NO. 1: PHYSICAL RISKS

Investors commit capital on the assumption of a stable climate and a predictable future. Now, financial returns are at risk from a rapid warming of the planet that raises the probability of uncertain outcomes. The physical impacts of a changing climate are already cropping up across America. In just one year, 2021, the country experienced devastating wildfires in California, multiple hurricanes in Texas, frequent flooding in Florida, and extreme heat waves in Oregon, destroying homes, businesses, and even entire communities. Investors are becoming aware that the physical impacts of climate change are putting financial assets at great risk.

The first risk for investors is in the nonlinearity of impact—as the climate changes, the value of assets is initially unaffected, but then a critical point is reached where the value of assets can quickly collapse. For example, buildings are designed to withstand floods up to a certain depth, below which little damage is done, above which the damage is immense. Similarly, crop yields decline marginally with a modest change in temperature until a threshold is reached, after which crops fail entirely.

The second risk for investors is timing. Physical risks from climate change, such as rising seas, storms, drought, and severe heat, will mostly occur decades in the future. But that does not mean asset values are unaffected today. Investments are valued by predicting cash flows over the life of an asset and applying a discount rate. Physical changes that might affect future cash flows increase the discount rate, which is a measure of uncertainty. In the era of climate change, investors must increasingly factor in the risks posed by physical changes and discount future cash flows accordingly.

The physical risk of climate change is the most obvious trend to affect asset values and investment returns, described in more detail in chapter 10. Other trends are more challenging to predict but are, for investors, of equal or even greater importance.

TREND NO. 2: TECHNOLOGICAL INNOVATION

Rapid innovation in low-carbon technologies is a second climate trend and, from the perspective of investors, the most important. For example, renewable solar and wind power is undercutting fossil fuels using advanced technologies to drive down the cost of energy to the point where competitors become insolvent. In 2020, for the first time ever, renewable energy generated more electricity in the United States than did coal, and five American coal companies filed bankruptcy in a single year.[1] Even the largest companies are experiencing setbacks. ExxonMobil, the longest serving member of the Dow Jones Industrial Index, was replaced because of declining market value.[2] After nearly 300 years, the fossil fuel industry in America is being superseded by cheaper, cleaner technologies.

In the automobile industry, the newest electric vehicles outperform gasoline-powered cars through quicker acceleration, better handling, and lower operating costs. Tesla was the first modern electric-vehicle company, incorporating innovative battery technologies and software into its automobiles. To the surprise of many investors, Tesla surpassed GM, Volkswagen, and Toyota to become the world's most valuable automobile company.[3] Technological innovation by one company has upended an entire sector, leaving traditional auto companies scrambling to play catch-up.

Perhaps the most surprising technological innovation is in the food and agriculture sector, where consumption of meat substitutes is threatening an industry that has existed since the dawn of human civilization. Investors see the potential for this innovation to disrupt an entire industry. Beyond Meat, a leading producer of plant-based meat substitutes, enjoyed the most successful initial public offering (IPO) of any American company since 2000, more than doubling in value before shares even began trading.[4]

Companies have always faced the risk of technological innovation from competitors. Now, innovators in electric power, transportation, and food are experiencing rapid growth because they offer consumers a superior product while addressing climate change. Investors are supporting this trend by providing growth capital, just as early investors did for James Watt and the inventors behind the Industrial Revolution. The result is an upending of asset values as nimble entrepreneurs challenge traditional industries. Section 2 of this book explores these technologies in greater detail.

TREND NO. 3: EVOLVING SOCIAL NORMS

Greta Thunberg, a high school student from Sweden (figure 2.1), became the face of a global movement of young people who are addressing climate change through their use of social media to profoundly alter attitudes toward action on climate change. Even *Forbes*, a publication aimed at America's wealthy, published an article titled "Why Greta Thunberg Is Totally Correct in Her Latest Climate Change Tweet." Thunberg and the millions of people who have attended her rallies represent a third trend influencing the flow of capital: evolving social norms.

Many Americans have long regarded climate change with skepticism. Support for government action reached a low in 2012, when only 25 percent said global warming should be a top priority for Congress. But that has changed, primarily due to the concerns of Gen Z and Millennials. By 2020, 52 percent of Americans surveyed said climate change should be a top priority for Congress,[5] and among American youth, support is much stronger for climate activism compared to that of older generations.[6]

Young people are often idealistic, but this generation's idealism is affecting businesses and investors in ways not seen before. In a tight labor market, employers are finding that sustainability matters to potential employees, and in highly competitive sectors it also matters to consumers. Younger Americans prefer to work for, and buy from, companies that act on climate change.

A key segment of today's youth is creating change in a different way, through their investments. Younger Americans are forecast to receive the greatest wealth transfer in history, an estimated $30 trillion, as the Baby Boomer generation passes on. Research finds that 86 percent of them are interested in sustainable investing, a rapid and dramatic change in social norms among an entire generation of investors.[7]

Evolving social norms are also changing the behavior of businesses, with companies across America pledging to reduce their greenhouse gas emissions. Amazon has committed to net-zero carbon,[8] as have dozens of other leading companies including Apple, Ford, and Starbucks.[9] Microsoft has gone one step further and committed to remove all the carbon the company has emitted since its founding.[10] Companies are taking action because it improves their competitive positioning, as explained in section 3 of this book.

Evolving social norms are changing individual and corporate behavior with respect to climate change. They are also forcing governments at the federal, state, and local levels to respond.

TREND NO. 4: GOVERNMENT ACTION

Greater awareness of the physical impacts of a changing climate, alongside evolving social norms, are at long last influencing voters to pressure governments to act. While international climate change agreements have failed to stem the rise of greenhouse gases, national and local governments have implemented a wide range of initiatives aimed at doing so. More than 100 countries have voluntarily set or are considering net-zero emissions targets,[11] including the United States.[12]

Many state and local governments are ahead of the federal government in pushing for aggressive regulations and incentives. New York State, for example, has pledged to eliminate all greenhouse gas emissions by mid-century, with incentives for renewable energy enticing developers to construct the largest offshore wind projects in the country. California's plan is for 100 percent clean energy by 2045, along with 14 other states that have passed legislation or executive action to move toward entirely clean electricity. Thirty states have created renewable portfolio standards, including Texas, Iowa, Montana, and many other states with Republican legislatures, and state initiatives are responsible for half of the growth in renewable energy generation.[13]

Politicians have learned that even modest policies go a long way, encouraging businesses to act and demonstrating to investors that emissions can be reduced, often at low cost. Even better, businesses and investors that commit to climate solutions provide cover for government officials to accelerate policy support. In this way, politicians and business leaders have learned that by working in concert, it becomes possible to enact policies that materially reduce greenhouse gas emissions, even in the absence of binding international agreements.

Government action to address climate change has been highly uneven and unstable since the first international agreement was negotiated in Rio de Janeiro in 1992. Despite that rocky path, the overall trend is clear, with government policies to tackle climate change increasingly ambitious at the national, state, and local levels. Governments are offering ever-greater financial and regulatory incentives for companies and projects that reduce greenhouse gas emissions and ever-greater penalties for those that do not.

A CONFLUENCE OF TRENDS, INFLUENCING INVESTORS

Climate change is a global tragedy of the commons that in a more perfect world would be addressed with a binding international agreement to eliminate greenhouse gas emissions. But that has not happened to date and may never happen quickly enough to address the climate challenge. Instead, a confluence of trends is driving capital away from fossil fuels and other polluting industries and toward companies and projects implementing climate solutions.

These trends are accelerating, providing the best and perhaps only remaining opportunity to avoid catastrophic climate change. Investors have a responsibility to understand the impact of these trends, both for their own financial benefit and for the greater good that comes from financing climate solutions. Like climate change, these trends have been slow to develop and poorly understood, but the momentum among investors is increasingly visible. As Hemingway famously wrote, wealth will be won and lost "gradually and then suddenly."[14]

"In the near future—and sooner than most anticipate—there will be a significant reallocation of capital."

FIGURE 3.1. Larry Fink, Founder, Chairman, and CEO, BlackRock, Inc. (*Source*: Kristoffer Tripplaar / Alamy Stock Photo)

Chapter Three

MOMENTUM

Scientists, academics, and policy makers have researched climate change for more than 30 years. But in the financial community, until recently, only a small number of investors studied the issue, and an even smaller number allocated capital with climate change in mind. Traditional institutional investors, managing most of the world's wealth, have been slow to recognize how climate change affects investment risks and returns. That is rapidly changing.

Larry Fink (figure 3.1), founder and CEO of BlackRock, the world's largest investment manager, published a letter in 2020 in which he wrote "there will be a significant reallocation of capital" from companies that ignore the perils of climate change. Fink wrote "climate change has become a defining factor in companies' long-term prospects. . . . I believe we are on the edge of a fundamental reshaping of finance."[1] BlackRock's announcement was driven by a recognition of the systemic climate trends that are affecting investors globally—changes in the physical environment, innovation in low-carbon technologies, evolving social norms, and expansion of government policies—not purely by some notion of social responsibility.

BlackRock is not the only company in the financial sector to recognize the implications of climate change for investors. Goldman Sachs plans to invest $750 billion in businesses responding to the problem, announcing "there is not only an urgent need to act, but also a powerful business and investing case to do so."[2] Among public fund investors, Marcie Frost, CEO of CalPERS, declared "we are integrating climate risk across our $388 billion portfolio."[3] And hedge fund managers also get it. Chris Hohn, founder and CEO of TCI, the world's most profitable large hedge fund in 2019 (assets: $30 billion), is clear in his position on the issue: "Investors don't need to wait on regulators who are asleep at the switch and unwilling or unable to regulate emissions properly. Investors have the power, and they have to use it."[4]

Systemic trends are upending investment flows, moving capital away from polluting industries and sectors at risk from climate change and toward companies and projects with climate solutions. The scale of investment is immense. Global investment in low-carbon solutions is forecast to total $125 trillion by 2050, with 70 percent of that capital coming from the private sector.[5] Successful investors are adapting their strategies in response.

Investment decisions have serious implications for humanity's ability to mitigate and adapt to climate change. Flows of investment capital will alter the trajectory of greenhouse gas emissions, slow climate change, and profoundly affect the future. This book explains the momentum in specific technologies, investment products, and sectors of the financial markets. Section 2 describes climate solutions—the technologies and business models in commercial operation today and in the not-too-distant future. Section 3 explains strategies used by investors in the era of climate change. Sections 4 and 5 examine opportunities and challenges for investors resulting from the trends described above, first for real assets and then for financial assets. Section 6 concludes with a discussion of investor timing, best practices, and why investment matters.

SECTION 2

Climate Change Solutions

Climate change cannot be eliminated. Even if emissions were immediately reduced to zero, the existing stock of greenhouse gases in the atmosphere would continue to warm the earth in the decades ahead. But catastrophic climate change, a warming of the planet by more than 1.5°C–2.0°C, can be avoided. This will require an unprecedented global reduction in greenhouse gas emissions, to near zero, by 2050.[1] After 250 years of steadily increasing emissions, a reversal and decline to zero in less than 30 years might seem improbable. In fact, it is both technically and economically possible.

Climate solutions already exist to dramatically reduce greenhouse gas emissions at modest cost. Goldman Sachs estimates that more than half of all greenhouse gases can be removed at an annual expenditure of $1 trillion using *current* technologies.[2] That's a large number, yet only 1 percent of current global GDP.[3] Importantly, these climate solutions are commercial, meaning that businesses can implement them at scale.

Investing in the era of climate change requires an understanding of climate change solutions prior to consideration of investment strategies. This section of the book considers only those climate solutions that are existing, scalable, and commercial. That does not mean other solutions for addressing climate change should be ignored or that noncommercial solutions are infeasible. But the focus must be on climate solutions that reduce the greatest volume of greenhouse gas emissions in the shortest amount of time using available commercial investment capital. The climate solutions described in this section do exactly that.

4. Renewable Energy
5. Electric Vehicles
6. Energy Storage
7. Green Hydrogen
8. Carbon Removal
9. Better Together

"I'd put my money on the sun and solar energy. What a source of power! I hope we don't have to wait until oil and coal run out before we tackle that."

FIGURE 4.1. Thomas Edison (*Source*: Alamy Stock Photo)

Chapter Four

RENEWABLE ENERGY

SOLAR POWER

Capturing energy from the sun has been a goal of scientists and engineers for more than a century.[1] Thomas Edison, the greatest inventor of his time, was attracted by the opportunity to harness the sun's power, as sunlight provides as much energy to the earth in 90 minutes as is consumed by every person on the planet in a year.[2] The theory for capturing solar energy was discovered by Albert Einstein in 1921, earning him a Nobel Prize in Physics for his discovery of the photoelectric effect.[3] The challenge has been to turn the potential of solar into practical commercial products. A century on, Edison's foresight and Einstein's brilliance have been proven entirely correct, with solar energy on the cusp of dominating the global power industry.

Early Applications of Solar Energy

The first practical application of solar energy was on the Vanguard 1 satellite launched by the United States in 1958. It was an auspicious start, as the solar panel operated continually for 7 years, substantially longer than the satellite's conventional batteries, which lasted only 20 days.[4] NASA went on to use solar panels on both satellites and spacecraft. While the use of solar panels flourished on space missions, they were considered far too expensive for most applications back on Earth.

Slowly but surely, solar energy spread, from space-based applications to off-grid locations in the Arctic, on drilling rigs, and on remote islands. The energy crisis of the late 1970s encouraged companies to develop better-performing solar products at lower cost. But growth in the solar sector was slow. By the end of the twentieth century, total installed photovoltaic systems globally had reached only 1 gigawatt (GW), equivalent to the electricity that could be produced by just one coal-fired or natural gas–fired power plant. Generating electricity from the sun was confined to a few niche applications for a simple reason: cost.

THE ECONOMICS OF ELECTRICITY

Electricity is a commodity, which means the electricity generated from coal-fired power plants is identical to what is produced by solar panels. Thus, the primary differentiating factor among different sources of electrical power generation is cost, and the cheapest form of power production becomes the preferred source.

Comparing the cost of different sources of electrical power generation poses a challenge. Coal-fired power plants are relatively inexpensive to build but require an enormous amount of coal to generate electricity. Solar farms are more expensive to build than coal plants but are less costly to run because the fuel from the sun is free. How to compare very different sources of electrical power?

The solution is a formula, the levelized cost of electricity (LCOE), which is a standard metric to compare the cost of electricity from different sources of generation. The LCOE provides an apples-to-apples comparison of cost. The LCOE for a power plant equals the cost of building and operating the plant and of sourcing the fuel divided by the electrical output forecast over the life of the plant, discounted at the cost of capital required to invest in the plant.

The LCOE is calculated over the projected lifetime of the power plant, typically 20–40 years, and is expressed in dollars per megawatt-hour (MWh) of electricity produced (or in dollars per kilowatt-hour [kWh], which is simply dollars per MWh divided by 1,000). LCOE allows for a cost comparison of very different sources of power using the same metric.

THE ECONOMICS OF SOLAR POWER

The raw materials required to produce electricity from sunlight are abundant and inexpensive to source. However, the manufacturing process is complex and costly. Photovoltaic (PV) systems are composed of solar panels or modules, each containing many solar cells. Solar energy, the generation of electricity from light, occurs when the solar cells convert light into electricity using semiconducting materials. In most solar cells, the semiconducting material is silicon, one of the most abundant materials on Earth and the same material used in computer chips. When light is absorbed by the silicon semiconductor, the energy in light photons moves electrons, which flow as electrical current through the solar cell along wire conductors. Not all the sunlight reaching the solar cell is converted to electricity. Conversion efficiency is the ratio between the useful output of an energy conversion device and the input. Most commercial PV panels have conversion efficiencies of approximately 20 percent.[5]

The cost of manufacturing solar PV panels is measured in dollars per watt. In the early days of the solar industry, when PV panels were used primarily by NASA on spacecraft, the cost per watt was more than $100. By 2000, that figure had declined to $5, an impressive reduction in cost.[6] Even this was not low enough for electricity from solar to have an LCOE competitive with other sources of electricity generation, except in niche applications. But each time the price of PV panels declined, the number of niche applications increased, and production of panels expanded.

As production of PV panels increased, costs further declined, creating a virtuous cycle. By 2020, the cost of manufacturing a solar PV panel had plummeted below $0.25 per watt, a 95 percent cost reduction in just two decades.[7] The decrease in panel prices resulted in a remarkable decline in LCOE for electricity generated using solar PV panels, making it competitive with other sources of electricity generation, including fossil fuels.[8] The learning curve was key to this transformation.

THE LEARNING CURVE APPLIED TO SOLAR POWER

The process by which costs are reduced as production volumes increase is called the *learning curve*, an economic concept that has significant implications for the growth of solar power. The theory underlying the learning curve is that a consistent improvement in performance or cost is possible through increased experience. Moore's law is the best-known application of the learning curve. Gordon Moore, the cofounder of Intel, forecast that the capacity of computer chips would double every 2 years, implying a learning curve of 40 percent, a forecast that held roughly true for 50 years.[9]

Importantly, the learning curve as it applies to advances in renewable energy is not time-bound like Moore's law, it is simply a way to demonstrate the rate of technological improvements relative to production volume regardless of how long those improvements take. The learning curve as applied to solar power measures the ability of the solar industry to reduce the cost of panels as the volume of production increases. The learning curve is calculated as the percentage drop in price for each doubling of cumulative production. Richard Swanson, the founder of solar panel manufacturer SunPower Corporation, observed that the cost of producing solar PV panels declines approximately 20 percent for each doubling in cumulative production.[10] This observation became known as Swanson's law (figure 4.2).[11]

Academic research has confirmed that Swanson's law is reasonably accurate, as learning curves for solar PV systems displayed a mean of 23.8 percent across multiple studies.[12] Those studies found that the learning curve works well for solar PV systems for several reasons, including economies of scale, improvements in conversion efficiency, and advances in manufacturing processes. Swanson's law has important implications for the LCOE of solar energy. As the solar industry grew, the price of solar panels declined, and because solar panels make up approximately half of the capital costs of a solar project, the overall LCOE declined as well.

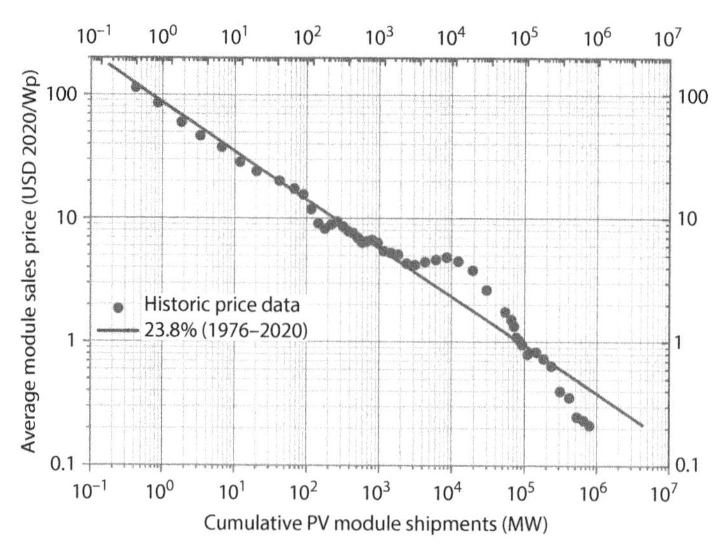

FIGURE 4.2. The learning curve for solar PV panels (Swanson's law). MW = megawatts, Wp = watt peak, the maximum electrical capacity a solar panel can generate.

Distributed Energy

Solar panels are a form of distributed power generation, as PV panels can be placed in almost any location and at any scale, from a single panel up to millions of connected panels. This means solar panels can be used to power everything from a handheld calculator to a house, an entire building, or even a city. Solar power can also be used off-grid, providing electricity in locations too remote or too costly for connection to the electrical grid. This stands in contrast to traditional centralized power generation, in which a large plant generates electricity that is transmitted to many users across the electrical grid. Distributed power generation has much greater flexibility in design than centralized power generation, a significant competitive advantage.

The more applications for solar power, the greater the demand for panels, increasing production and—because of the learning curve—driving down costs. As panel costs decline, the LCOE of solar power declines as well, further increasing demand, and so on, creating a virtuous cycle that is making solar power the lowest-cost source of electricity for most applications in most places on the planet.

Competitive Solar

Lazard, a multinational investment bank, publishes an annual analysis of the LCOE for renewables, nuclear, and fossil fuels on an *unsubsidized* basis. Lazard's analysis found that by 2020 the LCOE of solar power had declined 90 percent, making it competitive with all other sources of electricity.[13] More important, every year, solar power becomes even cheaper.

Solar technology is constantly improving. Trackers are now used on large solar projects, allowing solar panels to track the arc of the sun through the sky, increasing the electrical output by up to 30 percent.[14] Bifacial panels capture reflected sunlight from below, increasing electricity generation by an additional 9 percent.[15] Improvements in solar technology further reduce the LCOE of solar power, increasing demand, and the cycle begins anew, creating compound annual growth in the solar energy market of 42 percent every year for the past decade.[16] Wood Mackenzie, an energy consulting firm, predicts solar power will be "the cheapest source of new power in every US state and in Canada, China and 14 other nations."[17]

Challenges to Growth

The rapid growth of solar power faces two challenges: land and intermittency. Solar panels require approximately 5 acres of land to generate 1 MW of electricity to power 200 homes. Generating power for every home in America would require nearly 10,000 square miles of land.[18] That sounds like a lot of land but is less than one-quarter of the Mojave Desert. Even better, as a distributed form of power generation, solar panels can be placed on many surfaces. Solar could provide nearly 40 percent of U.S. power needs simply by installing panels on every rooftop.[19] It turns out that land is not an insurmountable problem, although it is often perceived as one. Intermittency is a greater problem.

Solar panels cannot generate electricity at night and generate little electricity when it is cloudy, making solar power an *intermittent* source of electricity. Intermittency is the drawback of solar power, as the modern economy relies on electricity to be available at all times. The solution to intermittency is to store excess electricity when it is sunny, but energy storage has been costly. Until now. As explained in chapter 6, battery energy storage is removing the final hurdle to widespread use of solar power.

The Future of Solar Power

Global energy firm BP forecasts that solar power will decline in cost a further 65 percent by 2050.[20] As the cost of solar power declines, demand from businesses will continue to grow. Amazon, a company well known for cutting costs, has committed to powering its entire business with 100 percent renewable energy by 2025, betting on the declining cost of solar to save money and lower greenhouse gas emissions.[21] And as solar projects increase in size, they will compete directly with large coal and natural gas plants, replacing fossil fuels. Government policies will further encourage the use of solar; for example, in 2020 California became the first state to require solar panels on new homes.[22]

For investors, the growth of solar power will create demand for capital on an unprecedented scale, a forecasted $4.2 trillion in new investment opportunities to 2050.[23] Sections 4 and 5 of this book explain how that capital will be invested. Solar represents one-half of the opportunity in renewable energy; the other half is in wind.

WIND POWER

Early American settlers erected windmills to pump water for irrigation and provide electricity in their homes.[24] But the use of windmills declined with the advent of the modern electrical grid, which provided reliable electricity at low cost. Today, farmers are once again installing windmills, this time to generate electricity for the entire country.

Wind turbines create power by capturing the energy in the wind. The amount of energy captured is a function of the speed of the wind and the area that is swept by the blades of the turbine. Wind speed is especially important, as the amount of energy in the wind is proportional to the wind speed cubed. For example, winds at 20 mph have *eight* times the energy of winds at 10 mph. Importantly, wind speeds are stronger at greater heights, meaning that a taller wind turbine is exposed to more energy than a shorter one.

Just as the energy in wind is proportional to speed, the area swept by the blades is proportional to the length of the wind turbine's blades. Area equals πr^2, where r is the length of the wind turbine's blade. A blade that is 12 meters long will sweep an area that is *four* times as large as a blade that is 6 meters long. Consequently, increasing the height of a wind turbine enough to double the blade length will quadruple the swept area of the blades. And a taller wind turbine will allow those blades to experience higher wind speeds. The combination of longer blades and taller turbines will result in a dramatic increase in energy captured.

The history of modern wind turbines is therefore one of increasingly bigger structures. The turbines developed in the 1980s had 15-meter rotor diameters (i.e., the diameter of the area swept by the blades), and each turbine generated 50 kW of power, enough for about 10 homes. By 2020, innovations in materials and technologies allowed for the construction of wind turbines with rotor diameters of 220 meters, generating 14,000 kW (or 14 MW) of power from each turbine, which is sufficient to power 2,800 homes (figure 4.3). The innovations and advances in technology that created bigger turbines resulted in a *280-fold* increase in power generation.[25]

The Economics of Wind Power

In the 1980s and 1990s, wind turbines were small, and the economics of wind farms was poor. The high capital costs of wind turbines meant that wind farms were expensive to build, and low power output meant they generated

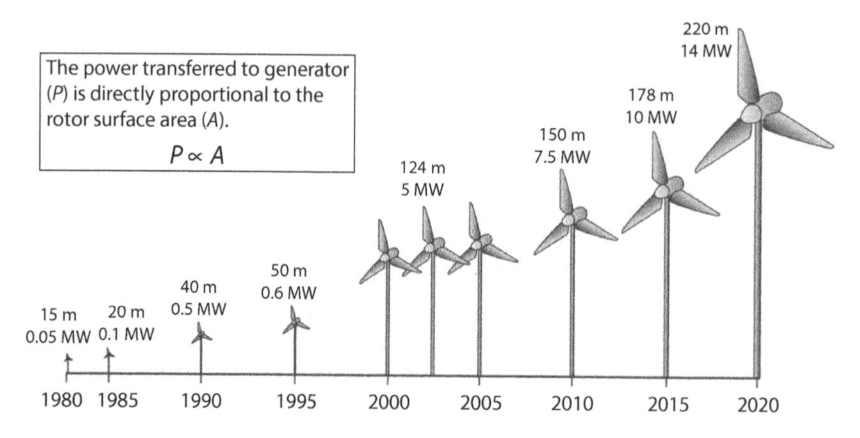

FIGURE 4.3. Relationship between wind turbine size and power

costly electricity. In comparison to electricity generated from coal, nuclear, and natural gas, early wind farms had a relatively high LCOE, making them uncompetitive with traditional forms of energy.

Government subsidies, especially in Europe, provided incentives for wind developers to build wind farms, creating demand for wind turbines from manufacturing companies. This created a tipping point, as government incentives compensated for the relatively poor economics of wind, allowing manufacturers and project developers to build wind projects, thereby creating expertise in the sector. As expertise improved, costs declined, and more wind projects were built.

Decreasing the LCOE of Wind Power

The physics behind generating energy from the wind led to the development of increasingly bigger and more efficient wind turbines. But when generating electricity, bigger is only better if it is also cheaper. The modern wind industry focused relentlessly on reducing the LCOE of wind power to improve competitiveness with other forms of electricity generation. It worked. For wind power, the LCOE has declined by more than 90 percent since the 1980s, making wind competitive with electricity generated by coal, nuclear, and natural gas.[26]

Manufacturers of wind turbines and developers of wind farms are constantly improving production techniques, applying new technologies, and taking advantage of economies of scale. And each time this occurs, wind energy companies learn how to reduce costs and increase output, thereby reducing the LCOE of wind power. The installed capacity of wind turbines has been doubling every four to five years, and as production has increased, costs have decreased. The results have been impressive. By 2020, wind farms were producing 8 percent of U.S. electricity, from near zero a decade earlier.[27]

Challenges to Wind Power

Stating the obvious, wind turbines only generate electricity when it is sufficiently windy to turn the turbine blades. On most wind turbines, this requires a minimum wind speed of 7 mph.[28] At low wind speeds, turbines do not generate any power, and even moderate winds do not optimize electricity generation. *Capacity factor* measures the electricity produced by a wind turbine divided by its maximum power capability. Overall, U.S. wind farms experience capacity factors averaging only 35 percent.[29]

Like solar power, wind power is intermittent, meaning it can only be produced when the wind is blowing. Conveniently, wind power provides a useful complement to solar generation, as winds are more powerful at night and also in the winter, when solar generation is lower.[30] This means wind can reduce the amount of energy storage required to maintain grid balance.[31] Intermittency and solutions to address it are explored in more depth in chapter 6.

Wind farms face additional hurdles, mostly related to siting and transmission of the electricity generated by turbines to the electrical grid. Wind turbines are popular with consumers seeking lower electricity costs and cleaner energy and among farmers and other landowners receiving annual payments for leasing land. Their neighbors are not always as keen. Not-in-my-backyard-ism (NIMBYism) is a chronic challenge to the development of wind farms. In California, long a leader in renewable energy, wind growth has slowed significantly in recent years as local opposition grows.[32]

Opponents of wind farms point to the fact that collisions with turbines result in up 500,000 bird deaths per year.[33] Sadly, that pales in comparison with bird deaths from collisions with buildings and from domestic cats. The American Bird Conservancy estimates that cats kill 2.4 billion birds, and a further 1 billion birds die from hitting buildings in the United States every year.[34] Even the Audubon Society has come out in support of wind power, pointing out that the risk to birds from climate change is far greater than from wind turbines.[35]

An additional siting challenge for wind farms is the need to transmit the electricity generated by the wind turbines to the electrical grid, where it is used by homeowners and businesses. The optimal locations for wind farms are where it is most windy; naturally, many of those places are far from where people live. In the United States, the windiest states are Nebraska, Kansas, and the Dakotas, none of which has significant population centers within the states or nearby.[36] As such, wind farms often require costly new transmission lines to move electricity from its source to where it will be used.

Despite these challenges, wind power has grown rapidly, and the sight of wind turbines is increasingly common. But the future of wind power might be out of sight, offshore.

Offshore Wind Power

Wind farms can also be placed offshore, where there is nearly unlimited space for new construction. Offshore wind is also steadier and more powerful than at onshore sites, so the capacity factor of offshore wind is 40–50 percent versus 35 percent onshore.[37] Offshore wind farms sometimes face less public resistance—if the turbines are more than 30 miles distant, they are no longer visible from shore—allowing for use of the tallest turbines on the market. As an example of the benefits of size in the offshore market, General Electric's massive new offshore wind turbine can produce enough electricity from a single rotation of its blades to power a household for 2 days.[38]

The first offshore wind projects were developed in Europe, primarily in the North Sea, with the first U.S. offshore wind project developed in 2016 off the coast of Rhode Island.[39] Multiple states are now moving forward with aggressive offshore wind project plans, particularly in the Northeast. New York State has set a goal of developing 9,000 MW of offshore wind capacity by 2035. The first project, named Empire Wind, is an 816 MW offshore wind farm located 20 miles from Long Island, in proximity to New York City. When complete in 2024, this single project will power more than 1 million New York homes.[40]

Offshore wind farms are, however, costly to construct and maintain in the unforgiving conditions of the ocean environment. Despite higher wind speeds and electricity generation, the LCOE of offshore wind is more than double the cost of onshore wind.[41] Even the very large Empire Wind project near New York City is costing twice that of the equivalent onshore wind farm.[42] Driving down offshore wind costs and lowering LCOE is essential for offshore wind to reach its potential. In addition to higher costs, growth in offshore wind faces a physical challenge: deep water.

Much of the U.S. coastline drops off steeply underwater, offering limited areas with depths of less than 200 feet, the maximum for offshore fixed turbines.[43] Solving this challenge has led to a recent innovation in the wind industry: the development of floating turbines, which can be anchored in water more than 3,000 feet deep.[44]

Floating Wind Farms

Wind turbines located far offshore benefit from stronger, more consistent winds. This raises the capacity factor, increasing electricity generation and reliability. The world's first floating wind farm achieved a 55 percent capacity factor, even higher than the 40–50 percent of offshore wind nearer the coastline.[45] Floating wind farms offer the potential for near limitless growth, but they suffer from the challenge of all new sources of power generation: high cost.

Floating wind turbines are nearly double the cost of fixed offshore wind turbines and four times the cost of onshore wind turbines.[46] Even with higher capacity factors, floating wind farms have an uncompetitive LCOE in comparison with other forms of power generation. But that may soon improve. Costs are likely to decline, ironically, by using the expertise of the fossil fuel industry. Oil companies are entering this sector of the wind business, capitalizing on the knowledge developed building floating platforms to extract oil and gas from beneath the seabed.

The Future of Wind Power

Wind power, the generation of electricity from wind turbines, has become a highly competitive source of energy. Relentless advances in technology combined with targeted government incentives has created a virtuous cycle of lower costs, growing demand, and a steadily declining LCOE for electricity. Global development of wind power will require an estimated $5.9 trillion in financing by 2050.[47] In the United States, offshore wind development alone will require $85 billion in investment capital by 2029.[48] Opportunities for investing in wind power are explained in sections 4 and 5.

NUCLEAR

Nuclear power is contentious. Supporters argue that nuclear offers a potentially limitless source of zero-carbon energy, while opponents point to safety risks and toxic waste. In the heated debate over nuclear energy, a key point is rarely heard: it is costly.

Only two new nuclear reactors have been completed in the United States in more than a quarter-century, both at the same facility in Tennessee.[49] In Georgia, the Vogtle nuclear power plant expansion, the only U.S. nuclear project currently under development, is years late and billions of dollars over budget.[50] Lazard estimates the LCOE for nuclear power is $163 per megawatt-hour, more than four times the LCOE of solar and wind and three times that of natural gas.[51] Given these high costs, the U.S. Energy Information Administration (EIA) projects a net decline in nuclear generating capacity between 2019 and 2050.[52] Professor Robert Rosner of the University of Chicago put it more bluntly: "New builds can't compete with renewables."[53] But new designs for smaller nuclear plants could change that.

A New Generation of Nuclear?

Several start-up companies are developing new technologies and innovations to reduce the cost of building and operating nuclear power plants. Many of these designs are for small modular reactors, or SMRs, mini nuclear reactors of up to 300 MW versus more than 1 GW for traditional reactors. Instead of using water and large cooling towers, these companies are using molten salt to cool the reactor at low pressure, with the goal of reducing cost and improving safety. According to Bill Gates, "there's a new generation [of nuclear power] that solves the economics, which has been the big, big problem. At the same time, it revolutionizes the safety."[54]

In theory, this new generation of nuclear power plants could offer a competitive source of power with no greenhouse gas emissions, like wind and solar, with the significant advantage of providing constant 24-hour baseload power. But cost remains a challenge.

In 2020, a leading developer of SMRs predicted it will have a levelized cost of energy of $65 per megawatt-hour when the first plant enters into operation in 2029.[55] In comparison, the EIA forecasts that by 2025, solar PV systems will have a levelized cost of $33 per megawatt-hour, and onshore wind will be at $34.[56] By the time this new generation of nuclear reactor becomes viable, the cost of wind and solar power will be significantly lower, and even advanced nuclear power may never catch up. M. V. Ramana, a professor at the University of British Columbia, summed up the challenge faced by nuclear power with a simple visual metaphor: "The basic idea, from a business perspective, is flawed. It's a kind of treadmill race, where one treadmill is going much faster."[57]

The Future of Nuclear Power

In spite of the challenges, investment in advanced nuclear power has increased dramatically. Private investors have committed more than $1 billion in venture capital to more than 50 start-ups,[58] and the Department of Energy has committed more than $10 billion in loan capital.[59] Bill Gates, one of the world's richest people, is so committed to advanced nuclear power that he became a founder and chairman of TerraPower, a leading nuclear power technology company.[60]

Advanced nuclear power remains a long shot but has one significant advantage over wind and solar power—it is reliable 24 hours a day, 365 days a year. Wind and solar are intermittent sources of power, available only when it is windy or sunny. The future of nuclear power, therefore, may have less to do with the cost of energy generation than with the cost of energy storage, described in chapter 6.

THE ENERGY TRANSITION

Renewable solar and wind generated only 11 percent of U.S. electricity in 2020,[61] but investment in solar and wind accounted for 78 percent of all *new* generating capacity.[62] This is unsurprising, as in many states solar and wind are the cheapest source of power. For example, a solar project in New Mexico contracted to sell electricity for $15 per megawatt-hour, a record low.[63] Compare this to the *operating cost* of a coal or natural gas facility, which ranges from $23 to $48 per megawatt-hour due to fuel costs, and the implication is clear: fossil fuels simply cannot compete with the declining cost of renewables.[64] Research by Professor Geoff Heal of Columbia Business School finds that the economic cost of entirely replacing fossil fuels with renewable energy in the U.S. electrical grid is near zero.[65] Affordable solar and wind power are giving rise to a global energy transition.

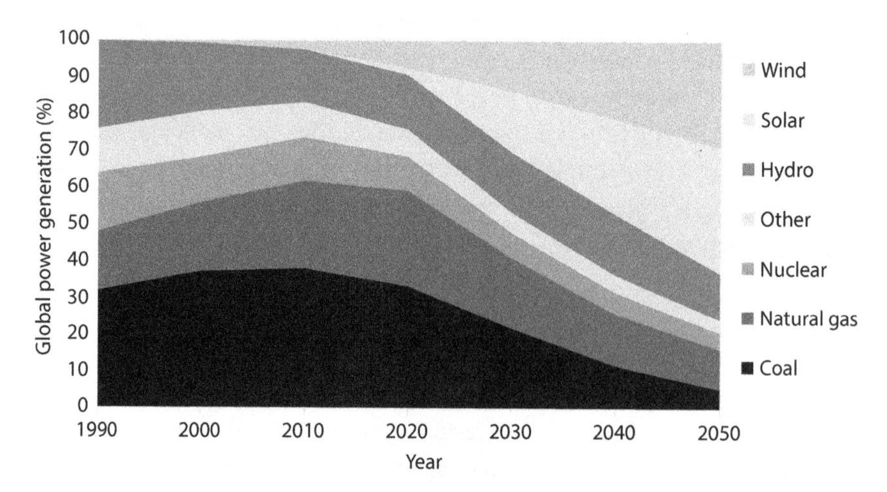

FIGURE 4.4. Global power generation. (*Source*: Author's estimates based on data from BNEF, IEA, and DNV)

The energy transition forecast in figure 4.4 is not wishful thinking by environmentalists; it simply reflects the increasingly competitive position of solar and wind technologies. The implications for investors in the fossil fuel industry are clear. Fatih Birol, executive director of the International Energy Agency, which has long focused primarily on the fossil fuel industry, summarized the situation: "The bitter truth is that real energy transitions are coming, and they are coming fast."[66]

The End of Coal

In 2020, Great Britain went several months without burning coal to generate electricity, the longest period without deriving energy from coal since 1790, the dawn of the Industrial Revolution.[67] The country is on track to entirely abandon coal by 2025.

In the United States, coal's share of electricity generation fell below 20 percent in 2020, a remarkable decline considering that coal generated more than 50 percent of American electricity only a decade before.[68] What accounts for the rapid decline in the use of coal? Utilities stopped buying coal, insurers stopped insuring it, and investors stopped investing in it.

Demand for coal has declined because of the better economics of generating electricity from natural gas, wind, and solar. The world's three largest reinsurance companies, Swiss Re, Munich Re, and Lloyds of London, have restricted insurance to the industry, making it difficult to operate.[69] And investors, seeing the coal industry become uncompetitive, are charging more for capital. In 2019, even with support from the Trump administration, eight U.S. coal-mining companies filed for bankruptcy, including Murray Energy, the largest private coal company.[70]

Natural Gas: A Bridge Fuel?

Natural gas is sometimes referred to as a "bridge fuel" between coal and renewable solar and wind for the generation of electricity. From the perspective of climate change, natural gas is indeed significantly less polluting than coal, emitting only half as much CO_2.[71] And from an economic perspective, natural gas is a low-cost source of electricity generation, with an LCOE of $63/MWh, higher than solar or wind, but with the added benefit of providing "dispatchable" power,[72] meaning it is always available. In 2020, natural gas generated 40 percent of U.S. electricity.[73] But the forecast for natural gas has declined markedly, suggesting the bridge to renewables may be a short one.

In a single week in July 2020, three American utilities announced plans to shutter coal plants and replace them with wind and solar farms, without any plans for natural gas as a bridge fuel.[74] The rationale is simple: energy storage is becoming cheap enough to be paired with wind and solar and outcompete not only coal, but natural gas as well. The announcement by one of those utilities, Tucson Electric Power, demonstrates that transition: "Even with the future planned retirement of 1,073MW of coal capacity and 225MW of natural gas capacity, TEP's Preferred Portfolio does not include the addition of any new fossil-fuel resources."[75]

Other Renewables

Generating electricity from rivers—hydropower—has historically complemented the role of coal in America. Hydropower is a low-cost, reliable, renewable power source, providing 7 percent of U.S. electricity generation in 2020.[76] There is further potential to refurbish and improve power generation in many smaller hydro facilities. But the role of hydropower is declining in the United States as the best sites for generating electricity are already taken and as environmentalists discover negative impacts from dam developments.

The EIA forecasts electricity generation from hydropower to decline to 5 percent of U.S. electricity generation in 2050.[77]

Energy generated from tidal, wave, biomass, biofuels, and geothermal contributes to the production of power, and each of these energy sources is renewable. Unfortunately, these technologies suffer from either high cost, limited growth opportunities, or relatively small markets. For example, geothermal power is cost-competitive with fossil fuels but only growing at 3 percent per year.[78] These renewable sources provide useful energy but are ill suited with current technologies to lead a global energy transition.

Solar and wind are the only sources of renewable energy that are both cost-competitive with fossil fuels *and* massively scalable in the United States and globally.

Winners and Losers in the Transition to Renewables

Consulting firm McKinsey & Company forecasts more than 50 percent of global power generation will come from renewable sources by 2035, the most significant transition in energy since the Industrial Revolution.[79] This is a win for avoiding catastrophic climate change. The energy transition is also an opportunity for investors as the global growth in solar and wind power already requires $300 billion per year in capital investment, forecast to grow to more than $1 trillion per year by 2035.[80]

For the fossil fuel sector, the shift in capital to renewables is driving up the cost of capital for new investments. Investment banks estimate that the hurdle rates on new oil and gas investments is 10–20 percent versus rates as low as 3–5 percent for wind and solar projects, prompting several large oil companies to transition their business models, with some spending half of their capital expenditures on low-carbon energy solutions.[81] Fossil fuel companies see the writing on the wall and have begun repositioning away from a shrinking industry and into growth sectors such as renewables.

The transition to renewable power still faces the challenge of intermittency, as solar and wind cannot provide electricity 24/7. Solving that challenge requires another climate solution, energy storage. But understanding the solution to energy storage first requires a detour to an entirely different climate solution: electric vehicles.

"The best performing car that *Consumer Reports* has ever tested."

FIGURE 5.1. Tesla's all-electric Model S, (Source: Wikimedia Commons)

Chapter Five

ELECTRIC VEHICLES

Henry Ford and other early automakers believed "there could be nothing new and worthwhile that did not run by electricity. It was to be the universal power."[1] By 1900, one-third of all vehicles on American roads were electric cars. New York City had a fleet of electric taxis, and the first-ever speeding ticket was issued to an electric taxi in Manhattan.[2] Even sports car legend Ferdinand Porsche developed an electric car.[3] The popularity of electric cars was self-evident; they were quieter, cleaner, and easier to operate than vehicles powered by internal combustion engines. Electric cars did not require a hand-crank to start and did not emit noxious-smelling exhaust. Convinced by the many advantages of the electric car, Thomas Edison and Henry Ford created a partnership in 1914 to design a low-cost electric car.

Edison and Ford well understood that electric vehicles (EVs) have several technical advantages over vehicles powered by internal combustion engines (ICEs). Foremost among these is that EVs convert energy into motion much more efficiently than ICE vehicles. ICEs burn gasoline to create heat, a process called combustion, which is then converted into mechanical energy to propel the vehicle. This process is inefficient, as only 17–21 percent of the energy stored in gasoline is converted to mechanical power. Electric motors, on the other hand, are highly efficient. By directly converting electrical energy into mechanical energy, an EV converts 59–62 percent of the electrical energy into powering the wheels.[4] Because the motor is more efficient, an electric vehicle requires one-third the energy of a gasoline-powered vehicle.

But the Achilles' heel of electric cars is the battery. All vehicles require a medium to store the energy used to propel them. Electric vehicles store energy in batteries, while vehicles powered by internal combustion engines store energy in gasoline. Batteries at the dawn of the automobile age were large and stored relatively little electricity, so energy density was low. Gasoline, by comparison, is energy dense. The same volume of gasoline can store more than 100 times as much energy as that in a lead-acid battery, the most common type of battery in Edison's day. ICE vehicles had a distinct advantage over EVs in the early 1900s, because gasoline was a much better way to store energy than the best batteries of the time, more than making up for the relative inefficiency of ICEs in converting that energy to mechanical power.

Edison and Ford were aware of the storage problem in EVs and tackled it by working to invent a battery with greater energy density. Henry Ford, quoted in the *New York Times* in January 1914, asserted: "The problem so far has been to build a storage battery of light weight which would operate for long distances without recharging. Mr. Edison has been experimenting with such a battery for some time."[5]

It did not work. Thomas Edison was the greatest inventor of his time, but he failed in his quest to build a better battery, ending the joint venture with Ford, and along with it the potential of EVs.[6] It was nearly 100 years before EVs would become commercially viable.

INNOVATION IN THE AUTO INDUSTRY

The oil crisis of the 1970s spurred automakers to reconsider electric vehicles as an alternative to gasoline-powered internal combustion engines. Engineers revisited the dilemma Edison and Ford had faced and set out to design a battery that was lightweight yet energy dense, capable of storing enough energy to power a modern car. Again, they failed.

In 1990, California enacted a program to promote the use of zero-emission vehicles, which created an incentive for automakers to design and produce electric cars. The draw of selling cars in California convinced the major automakers to once again invest research and development dollars to create a successful electric vehicle. General Motors (GM) introduced the first mass-produced all-electric car, the EV1, in 1996, with a lead-acid battery weighing 1,175 pounds and a range of 60 miles per charge.[7] But the vehicle was expensive to produce, and the limited driving range was unappealing to consumers. GM canceled the EV1 program after only a few years and destroyed the cars it had built, claiming the vehicle was not commercially viable. Regrettably for GM, the company's ill-fated foray into EVs inspired the documentary *Who Killed the Electric Car?*, which suggested a conspiracy between the auto and oil companies to prevent the development of electric vehicles.

While a conspiracy is unlikely—GM vehemently denied it—the company's decision to destroy every EV1 produced, by crushing them, was surely one of the more poorly considered PR plans in business history. More important, despite significant research and development, GM, Chrysler, Ford, Toyota, Honda, and other leading automotive companies failed to design an electric car that would travel as far or as quickly as gasoline-powered vehicles. Consumers were unimpressed with the electric cars that were offered, and every electric model designed during this period was eventually withdrawn from the market.

In 2003, a new automobile company was formed, Tesla Inc., named after Nikola Tesla, inventor of the AC electric-induction motor. Elon Musk, Tesla's CEO, announced "The Secret Tesla Motors Master Plan (just between you and me)":

> Build sports car
> Use that money to build an affordable car
> Use *that* money to build an even more affordable car[8]

Musk's announcement was tongue-in-cheek, but Tesla's business plan was very serious and very innovative in several ways. First, Tesla targeted the sports car and luxury high-end of the car market for its initial products. Previously, automobile companies had focused on the economy or low-end of the market to sell EVs. Second, Tesla focused on vehicle performance as the primary selling feature, as opposed to environmental benefits. Third, and most important, Tesla recognized that the solution to Edison's battery challenge was not to invent a new kind of car battery, but to take advantage of the extraordinary progress already achieved in batteries for an entirely different market, consumer electronics and mobile phones.

Instead of developing a new special-purpose car battery, the engineers at Tesla designed a battery pack composed of 7,000 lithium-ion batteries, the same batteries used in consumer electronics. The battery pack took advantage of the best characteristics of lithium-ion batteries, which are relatively small, lightweight, and energy dense at a reasonable cost. Rather than spend time and money inventing a new battery, the company's engineers focused on developing the software to control the battery pack and designed a proprietary powertrain connecting the batteries to an all-electric motor.

The first car that Tesla launched, the Roadster, was expensive, attractive, and fast. Electric motors produce significantly more torque—the rotational force from the motor to power the wheels—than an internal combustion engine, allowing for greater acceleration. Tesla's Roadster accelerated from zero to 60 miles per hour in a remarkable 3.9 seconds. In 2008, *MotorTrend* magazine reviewed the car, concluding that it is "profoundly humbling to just about any rumbling Ferrari or Porsche that makes the mistake of pulling up next to a silent, 105-mpg[9] Tesla Roadster at a stoplight."[10]

OVERCOMING CHALLENGES

The Roadster proved that an electric car could perform competitively against high-end sports cars. Tesla had successfully completed the first phase of its master plan. The second phase, to build a competitive high-end luxury car, was significantly more ambitious, requiring additional innovation. Tesla's next car, named the Model S (figure 5.1), was a sedan designed to compete with the BMW 5 Series and similar luxury cars. To be successful, Tesla had to overcome three challenges simultaneously: range anxiety, performance, and cost.

"Range anxiety" is a driver's fear that the battery in an electric car will run out of power before reaching its destination or a place to recharge. Prior to Tesla, commercial EVs had never exceeded a range of 60 miles, while gasoline-powered cars typically average well over 300 miles on a single tank and can be conveniently refueled in five minutes at any gas station. In theory, electric cars can be recharged using a regular electrical outlet, but in practice recharging from a 110-volt household outlet requires several hours or longer. Tesla tackled range anxiety in two ways. The company made the battery pack significantly larger in the Model S, expanding the range between charges to between 210 and 300 miles, and began installing *superchargers*, high-voltage charge stations, across the country. Superchargers allowed Tesla drivers to recharge their batteries to a range of 150 miles in approximately 30 minutes.

The expanded battery pack in the Model S was heavy, unsurprisingly, given the relatively low energy density of batteries. The additional weight risked affecting the vehicle's performance. Tesla addressed this by placing the battery pack on the floor of the vehicle, between the axles, which improved handling by lowering the car's center of gravity. The company also increased the engine's power. The result was astonishing. The Model S could accelerate faster than its competitors, going from 0 to 60 mph in less than 4 seconds, and handled better. *MotorTrend* magazine rated it "Car of the Year" in 2013.

Cost remained the single biggest challenge for Tesla. The battery pack in the Model S was estimated to cost $15,000, accounting for 25 percent of the total manufacturing cost of the car.[11] Government incentives offset some of these additional costs, especially the federal tax credit of up to $7,500 for the purchase of all-electric and plug-in hybrid vehicles.[12] Although the Model S had a higher initial cost than comparable vehicles, running costs were lower. Analysts at investment bank Credit Suisse estimated that the average owner of a Model S would spend $34 a month on fuel costs, compared with up to $175 a month for the equivalent gasoline-powered mid-size luxury sedan, and significantly lower servicing costs because of the simplicity of the electric motor.[13]

INNOVATION BEYOND THE BATTERY

Musk believed that Tesla's all-electric cars created the opportunity for further innovation in the automobile industry. Electric vehicles are much simpler than vehicles powered by internal combustion engines because they have fewer moving parts. With their simple design, EVs require fewer repairs and less servicing, a significant value to consumers. Analysts at J.P. Morgan found that "EVs have 20 moving parts compared to as many as 2,000 in an ICE, dramatically reducing service costs and increasing the longevity of the vehicle," and concluded that the running costs for an electric vehicle can be 90 percent lower than for a gasoline-powered one.[14]

This design advantage led Tesla to make the innovative decision to sell its vehicles directly to consumers through a network of company-owned stores, instead of the franchise dealerships used by other mass-market automobile companies. Musk believed that franchise dealers have a fundamental conflict of interest, as the typical dealer earns significantly more on replacement parts and service than from new car sales.[15] EVs like the Model S, however, are unlikely to require as many repairs as automobiles with ICEs, potentially upending the franchise dealer business model. Tesla's direct distribution model is designed to minimize service costs for consumers over the life of the vehicle, thereby further improving the cost-competitiveness of EVs in comparison with ICE vehicles.

"THE FEEL OF THE WHEEL SEALS THE DEAL"

This expression has been used by car salesmen for decades to describe the effect on drivers of test-driving a new car. Little did they know it would apply to EVs. The Model S was an immediate success, and drivers raved about the car's performance and handling. Launched in 2012, it quickly became the top-selling all-electric car globally. In 2015, *Car and Driver* named the Model S the "Car of the Century."

The final step in Tesla's original master plan, "build an even more affordable car," was realized in 2016 with the release of the Model 3, a mid-sized all-electric sedan with a range of 263 miles, designed to compete with the BMW 3 Series and similar vehicles. Within a week, Tesla received 325,000 preorders with a $1,000 deposit on each order, representing sales of more than $14 billion.[16] By 2021, the Tesla Model 3 had become the world's best-selling EV in history.[17]

COMPETITION

The surprising success of Tesla's electric vehicles spurred incumbent automobile companies to design and launch their own EVs. But the leading automobile companies were tepid in their initial response to Tesla, committing relatively small development budgets and expertise. Unlike Tesla, the largest car companies did not vertically integrate into manufacturing batteries, the most important technology in an EV, and they adapted existing ICE vehicle platforms to EVs instead of investing to develop a dedicated EV product.[18] To make matters worse, their existing networks of car dealers were less than enthusiastic about selling EVs that require few repairs when dealers earn 49 percent of gross profit from servicing vehicles.[19] The results for the world's leading automobile companies reflected those decisions; in 2021, twenty car companies sold EVs in the United States, yet Tesla's market share was an astonishing 72 percent.[20]

THE ELECTRIC VEHICLE MOVES INTO POLE POSITION

Electric vehicles face three significant challenges to replacing gasoline-powered ICE vehicles: charge spots, charge time, and cost. American drivers of ICE vehicles can refuel at any of 115,000 gas stations.[21] In contrast, drivers of EVs in 2021 had only 43,000 charging stations to choose from, and not all of those stations are compatible with every model of electric vehicle.[22] Charge times are also significantly longer than the time it takes to refuel a gasoline tank, a minimum of 30 minutes and a maximum of up to several hours.[23] And then there is the cost of purchasing an EV, which is higher than the cost of an ICE vehicle. But those challenges to EV sales are rapidly fading.

The price of lithium-ion battery packs, the costliest component in an EV, has declined from more than $1,100/kWh in 2010 to $137/kWh a decade later, with further cost declines expected in the years ahead.[24] Bloomberg New Energy Finance predicts cost parity for EVs and ICE vehicles will arrive in 2024, projecting "there will be no price difference and no EV sticker shock."[25] At that point, the significantly lower cost of refueling and servicing an EV will become an attractive proposition to new car buyers. Meanwhile, the availability of charge spots is growing rapidly, even as most EV owners charge their vehicles at home or work, and charge times are forecast to decline with advances in battery technologies.[26]

THE RACE IS ON

Automobile companies must plan 5 years ahead, the time it takes to develop a new model. The future they see is all-electric. While EVs face challenges today, the trends all point in the same direction: the EV is rapidly becoming a cheaper, faster, better product than automobiles using the internal combustion engine. Automakers also recognize that EVs will be selected for autonomous vehicles when self-driving technology becomes widespread, as EVs are more efficient for fleet management.[27] Facing these trends, incumbent automobile companies are rapidly abandoning their gasoline-powered vehicles and developing new EV models, often on dedicated platforms.

In 2021, GM, the largest U.S. automobile company, announced that it will manufacture and sell only EVs by 2035, spending $27 billion over 5 years to launch 30 electric vehicles.[28] Shortly after GM's announcement, the CEO of Volkswagen, the world's largest car company, had this to say about electric vehicles: "Let me begin with the obvious: e-mobility has won the race."[29] Analysts at UBS bank predict EVs will take 100 percent of the new car market by 2040.[30]

The transition from gasoline-powered to electric vehicles affects several major sectors of the economy: auto companies, parts suppliers, auto dealers, oil companies, and electric utilities. The implications for investors are examined in section 5 of this book.

Electric vehicles represent an important climate change solution as transportation accounts for 28 percent of US CO_2 emissions.[31] EVs require one-third the energy of ICE vehicles and can be powered by zero-emissions solar and wind. But EVs have another, lesser known, role in tackling climate change. The rapid growth in EVs is driving down the cost of lithium-ion batteries, providing an energy storage solution to the intermittency challenge of renewable solar and wind.

"The world's overloaded power grids are on the cusp of a megabattery boom."

—BLOOMBERG

FIGURE 6.1. Power transmission lines. (Source: Wikimedia Commons)

Chapter Six

ENERGY STORAGE

Renewable wind and solar power are cheap, abundant, and produce zero-emissions electricity. But they have one significant drawback: intermittency. Wind turbines cannot generate power on calm days, and solar panels cannot generate power at night or on cloudy days. Given the intermittency inherent in solar and wind generation, there is a growing need for energy storage to balance the demand and the supply of power. Understanding energy storage first requires an explanation of how electricity is managed on the grid.

THE LOAD PROFILE

Electricity is consumed in varying amounts throughout the day, described by a *load profile*, a graph of demand for electricity in a 24-hour day.[1] Electricity demand is typically lowest during the night, when most people are sleeping and businesses are closed. During the day, the load profile increases, especially in warmer regions with air conditioning. The load profile typically reaches a peak, referred to as *peak load*, in the late afternoon or early evening, when air temperatures reach a daily high and workers return home to switch on air conditioning, lights, appliances, and other electrical devices.

Managing the load profile is an essential task of a modern electrical grid, as consumers and businesses expect electricity to be continually available. Fluctuating demand for electricity requires the grid operator to supply

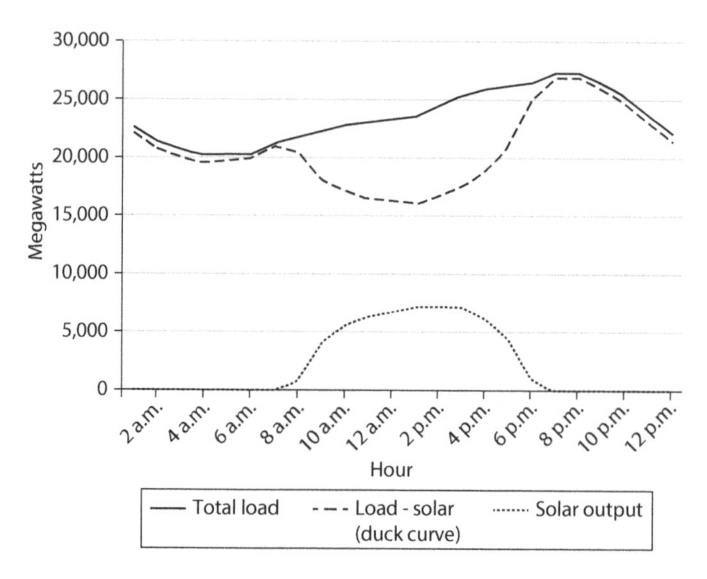

FIGURE 6.2. California's load curve (*Source*: Wikimedia Commons)

enough electricity to meet peak load without oversupplying electricity when demand falls. Managing the grid becomes harder when the generation of electricity is from intermittent sources such as wind and solar power.

The challenge posed by the intermittency of wind and solar power is already evident in California, where generation of electricity by renewables has affected the entire state's load profile for electricity. Figure 6.2 is the load profile for California.[2] Demand for electricity, as shown by the line labeled Total Load, hits a low at about 5:00 A.M., then slowly increases throughout the day, peaking at about 7:00 P.M. The problem that has emerged is that generation of electricity from solar power, as shown by the Solar Output line, peaks around 2:00 p.m. This distorts the net demand curve—the difference between overall demand and solar power output—to the line labeled Load – Solar. This distorted load profile vaguely has the profile of a duck, hence it is referred to in the utility industry as the "duck curve."

The grid operator must ensure that the supply of electricity precisely meets the demand curve throughout every 24-hour day, a challenge made increasingly difficult by the growing supply of electricity from wind turbines and solar panels. The duck curve in California is worsening every year with the addition of renewables. California is a leader in renewable energy, but other states are rapidly catching up and experiencing the same challenge balancing the load profile.

Solutions to the duck curve include demand response, peaker plants, and energy storage.

Demand Response

Utilities encourage customers to reduce demand for electricity during peak load periods, a tactic called *demand response*. Utilities do this by offering a financial incentive to customers, for example by reducing the utility bill of customers who install sensors that automatically increase the temperature setting on air conditioning during especially hot summer days. Demand response can assist utilities in balancing the load profile, but it cannot solve the problem of intermittent wind and solar power at scale.

Peaker Plants

Across the United States there are approximately 1,000 peaker plants that use natural gas to supply electricity on an as-needed basis.[3] Peaker plants have the advantage of being "dispatchable," meaning they can be rapidly turned on and off. But they are costly. Investment bank Lazard calculates the average natural gas peaker plant levelized cost of electricity (LCOE) at $175/MWh, five times the cost of power from renewables.[4] Costly peaker plants solve the challenge of balancing the load profile, but at great cost.

Energy Storage

More than 96 percent of global grid-scale storage capacity is in the form of pumped hydro storage, a technology developed more than 100 years ago.[5] Pumped hydro storage uses electricity to pump water from a lower-elevation reservoir to a higher one. When electricity is needed, the process is reversed, and stored water from the higher reservoir is released through turbines, which generates electricity. Pumped storage hydro is a reliable, dispatchable system for producing electricity on demand, with a round-trip efficiency that converts 70–75 percent of the electricity used to pump water back to the grid when needed.[6]

Unfortunately, pumped hydro is a poor solution to the energy storage challenge resulting from the rapid growth in solar and wind power generation. Pumped hydro storage is only feasible where the geology provides for two large water reservoirs conveniently separated by significant elevation, which is present in very few locations. Even more prohibitive, pumped hydro storage is costly to construct, and costs are not forecast to decline given the extensive engineering and construction required to build pumped hydro storage sites.[7]

Many other technologies exist for storing electricity on the grid, including compressed air, multiple battery technologies, and even flywheels. Among these technologies, lithium-ion batteries have the advantage of being extremely efficient, exhibiting round-trip efficiency of 92–93 percent. Conveniently, manufacturing costs of lithium-ion batteries are rapidly declining, driven by demand in an entirely different sector.

MEANWHILE IN THE AUTOMOBILE MARKET . . .

Tesla became the first American automobile company since the Second World War to successfully enter the mass market and the first to offer exclusively electric vehicles (EVs). But the success of the Model S created a problem for Elon Musk and his team. Tesla had taken advantage of the declining cost of lithium-ion batteries for laptops and mobile phones, following growth in those sectors to lower costs. However, Tesla's demand for lithium-ion batteries to build vehicles outstripped demand from laptops and mobile phones. It was estimated that production of the Tesla Model S in 2014 accounted for 40 percent of all lithium-ion batteries manufactured globally.[8] This prompted Musk to try another innovation, the Gigafactory, with the objective of reducing the manufacturing costs of lithium-ion batteries.

BIRTH OF THE GIGAFACTORY

In 2014, Tesla broke ground in Sparks, Nevada, on what it calls the Gigafactory, the world's biggest lithium-ion battery factory. It is difficult to overstate its size. When fully complete, it will be the largest building in the world,[9]designed to take advantage of manufacturing scale to lower the price of Tesla's battery packs by 30 percent.[10] Jessika Trancik, a professor at MIT, found that "lithium-ion battery technologies have improved in terms of their costs at rates that are comparable to solar energy technology," estimating that costs decline 20–31 percent for every doubling in market size.[11]

This creates a virtuous circle for Tesla. Lower battery prices make Tesla's cars more cost-competitive, increasing demand for them, increasing the production of batteries, decreasing the cost of manufacturing, and thus again driving up demand. Tesla's relentless focus on batteries drove costs down to $142/kWh in 2021,[12] less than half the cost when the Model S was launched. Elon Musk predicts a further 56 percent decline over the coming three years.[13] This is good news for Tesla. And great news for renewable wind and solar power.

DECAPITATING THE DUCK

Taking advantage of the rapidly declining cost of lithium-ion batteries, battery energy storage systems (BESSs) are replacing natural gas peaker plants to balance the load profile of the electrical grid. This solution is referred to as "decapitating the duck," as lithium-ion batteries store excess electricity generated from wind and solar and feed it back to the grid in the evening and at other peak times of the day, eliminating the bulge in the load profile curve. Not only are BESSs less costly than peaker plants, but lithium-ion batteries also have faster response times and can be instantly ramped up and down to balance the load on the electrical grid.

Vistra, a leading American utility, built the world's largest BESS in a retired natural gas power station in California. The Moss Landing project has 300 MW of power for 4 hours, providing 1,200 MWh of storage capacity, enough power for 225,000 homes.[14] The project can be expanded by a factor of 5 to 1,500 MW at a future date.[15] Vistra is not alone with this strategy. The U.S. Department of Energy forecasts 27 percent compound annual growth in grid storage through 2030.[16]

The extraordinary importance of lithium-ion batteries was recognized in 2019 when the Nobel Committee awarded the Nobel Prize in Chemistry to three scientists, announcing: "This lightweight, rechargeable and powerful battery is now used in everything from mobile phones to laptops and electric vehicles. It can also store significant amounts of energy from solar and wind power, making possible a fossil fuel-free society."[17]

Ever cheaper lithium-ion batteries will solve the short-term energy storage problem associated with the daily load profile. Unfortunately, batteries are ineffective at providing long-term energy storage to handle multi-day periods of low wind and sun. Addressing the intermittency challenge of renewables will include two additional solutions.

BACKUP ENERGY: V2G

In February 2021, winter storms pummeled Texas and knocked out power generation in much of the state, leaving 3 million homeowners in the dark. Electric vehicles can, in the future, prevent those blackouts.

Electric vehicles have the capability to send electricity back to the grid when they are not being used, a process called vehicle-to-grid (V2G) or bidirectional charging. In this way, electric vehicles are positioned to provide backup energy when the grid becomes overloaded. Ford's F-150 Lightning pick-up truck is the first electric vehicle in the United States to offer V2G capability.[18] V2G can provide significant backup energy; for example, Ford's F-150 has batteries capable of powering the average American home for up to 10 days.

Tesla sells a home battery system called the Powerwall, which uses the same lithium-ion batteries found in its automobiles. Like V2G, home battery systems can reduce grid instability. In California, homeowners with a Tesla Powerwall battery are participating in an experimental "virtual power plant" to send electricity back to the grid during periods of high demand, helping to reduce power outages and potentially generating additional revenue for Powerwall owners.[19]

V2G and home battery systems are a viable solution for backup energy storage, solving the problem of blackouts during periods of severe weather or other dislocations on the electrical grid. But they cannot provide long-term energy storage capable of maintaining reserve capacity for many days or weeks, a challenge requiring a different climate solution.

LONG-TERM STORAGE AND THE ENERGY TRANSITION

The global energy transition to wind and solar power will create a growing need for long-term storage to ensure electrical grid stability. A report by the University of California, Berkeley, estimates that a U.S. grid using 90 percent renewable energy would require 150 GW of storage capacity rated for 4 hours (i.e., 600 GWh),[20] a more than 100-fold increase on current capacity.[21] BESS projects using batteries will meet some of that demand, but batteries are a poor solution for multi-day energy storage.

Venture capital investors are financing engineers who are developing a wide range of multi-day energy storage products, including new battery designs, capacitors, flywheels, pumped air, and gravity systems. These technologies hold promise, but all face considerable technical and commercial hurdles. Fortunately, there exists an entirely different climate solution that can provide long-term, scalable energy storage: green hydrogen.

"Hydrogen will become, 30 years from now, like oil is today."

—SEIFI GHASEMI, CEO, AIR PRODUCTS & CHEMICALS, INC.

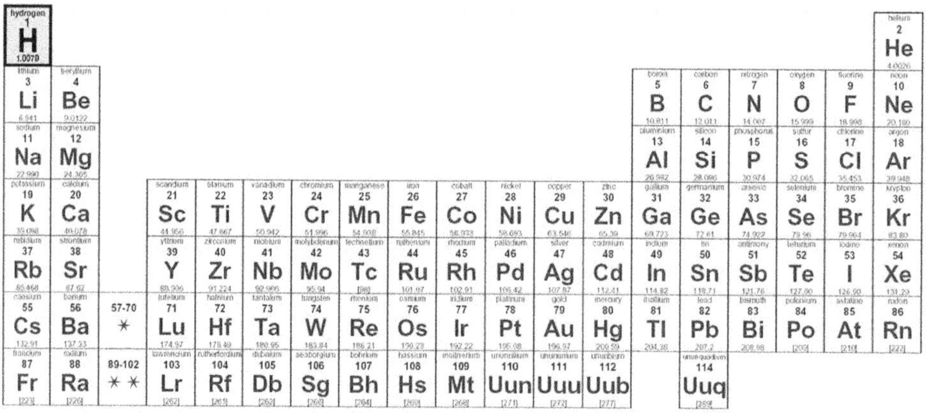

hydrogen 1 **H** 1.0079																		helium 2 He 4.0026
lithium 3 Li 6.941	beryllium 4 Be 9.0122												boron 5 B 10.811	carbon 6 C 12.011	nitrogen 7 N 14.007	oxygen 8 O 15.999	fluorine 9 F 18.998	neon 10 Ne 20.180
sodium 11 Na 22.990	magnesium 12 Mg 24.305												aluminium 13 Al 26.982	silicon 14 Si 28.086	phosphorus 15 P 30.974	sulfur 16 S 32.065	chlorine 17 Cl 35.453	argon 18 Ar 39.948
potassium 19 K 39.098	calcium 20 Ca 40.078		scandium 21 Sc 44.956	titanium 22 Ti 47.867	vanadium 23 V 50.942	chromium 24 Cr 51.996	manganese 25 Mn 54.938	iron 26 Fe 55.845	cobalt 27 Co 58.933	nickel 28 Ni 58.693	copper 29 Cu 63.546	zinc 30 Zn 65.39	gallium 31 Ga 69.723	germanium 32 Ge 72.61	arsenic 33 As 74.922	selenium 34 Se 78.96	bromine 35 Br 79.904	krypton 36 Kr 83.80
rubidium 37 Rb 85.468	strontium 38 Sr 87.62		yttrium 39 Y 88.906	zirconium 40 Zr 91.224	niobium 41 Nb 92.906	molybdenum 42 Mo 95.94	technetium 43 Tc [98]	ruthenium 44 Ru 101.07	rhodium 45 Rh 102.91	palladium 46 Pd 106.42	silver 47 Ag 107.87	cadmium 48 Cd 112.41	indium 49 In 114.82	tin 50 Sn 118.71	antimony 51 Sb 121.76	tellurium 52 Te 127.60	iodine 53 I 126.90	xenon 54 Xe 131.29
caesium 55 Cs 132.91	barium 56 Ba 137.33	57-70 *	lutetium 71 Lu 174.97	hafnium 72 Hf 178.49	tantalum 73 Ta 180.95	tungsten 74 W 183.84	rhenium 75 Re 186.21	osmium 76 Os 190.23	iridium 77 Ir 192.22	platinum 78 Pt 195.08	gold 79 Au 196.97	mercury 80 Hg 200.59	thallium 81 Tl 204.38	lead 82 Pb 207.2	bismuth 83 Bi 208.98	polonium 84 Po [209]	astatine 85 At [210]	radon 86 Rn [222]
francium 87 Fr [223]	radium 88 Ra [226]	89-102 * *	lawrencium 103 Lr [262]	rutherfordium 104 Rf [261]	dubnium 105 Db [262]	seaborgium 106 Sg [266]	bohrium 107 Bh [264]	hassium 108 Hs [269]	meitnerium 109 Mt [268]	ununnilium 110 Uun [271]	unununium 111 Uuu [272]	ununbium 112 Uub [277]		ununquadium 114 Uuq [289]				

FIGURE 7.1. The periodic table, with hydrogen emphasized (Source: Wikimedia Commons)

Chapter Seven

GREEN HYDROGEN

Renewable solar and wind power are the fastest-growing sources of new power generation on the planet, and declining costs will continue to accelerate the energy transition from fossil fuels to renewables. Simultaneously, the declining cost of batteries is solving the challenge of solar and wind power intermittency, providing short-term energy storage to balance the grid's load profile. But there are several sectors of the economy that are poorly suited to electrification. Air travel, long-distance ocean freight, heavy trucks, fertilizer, and industrial processes such as production of steel cannot easily be converted to electricity. And long-term energy storage for the electrical grid still relies on fossil fuel–powered peaker plants. Hydrogen, specifically "green hydrogen," is an attractive alternative, a potentially unlimited source of energy that does not emit greenhouse gases.

Hydrogen has long been used as a source of energy. Early in the twentieth century, hydrogen blimps provided the first transatlantic air travel, and hydrogen is used today in a range of industrial applications, from oil refining to fertilizers. Hydrogen is an appealing fuel because it is the lightest element (figure 7.1), providing twice as much energy per unit mass as oil or natural gas. Unfortunately, hydrogen is a challenging fuel to use as it is highly combustible and needs dedicated infrastructure for distribution. Because hydrogen is so light, the energy density per unit volume is very low, requiring high-pressure systems to liquify hydrogen for transport. But the greatest challenge to hydrogen is cost.

Hydrogen is the most abundant element in the universe yet occurs naturally on Earth only in compound form with other elements—hydrogen combined with oxygen is water, and hydrogen combined with carbon forms the hydrocarbons found in fossil fuels. To use hydrogen as a fuel, it must first be separated from other compounds. Steam methane reforming is the most common method for production of hydrogen, in which natural gas is used as the source of methane in the process. This is commonly referred to as "gray hydrogen." Unfortunately, production of hydrogen using steam methane reforming does not mitigate climate change, as the natural gas used in the process emits CO_2.

Hydrogen can also be produced using technologies to capture and store CO_2 emissions, a climate mitigation technique described in the next chapter. Production of hydrogen in this way avoids emissions of greenhouse gases and is referred to as "blue hydrogen"; however, it is significantly more costly than gray hydrogen because of the expense of capturing and storing CO_2. From a climate change perspective, blue hydrogen is better than gray hydrogen, but from an investor perspective it is uneconomical without large government subsidies or other incentives. Fortunately, there is another process for production of hydrogen that has the potential for both low cost and low emissions.

GREEN HYDROGEN

Hydrogen can be produced by splitting water into its respective elements, oxygen and hydrogen. The technology to do this, an electrolyzer, uses an electrical current to separate water molecules. Electrolyzers require significant energy to operate, which is both costly and polluting if the power is generated from the burning of fossil fuels. Fortunately, the rapid growth in renewable solar and wind power creates an opportunity to produce hydrogen using zero-emissions electricity, and to do so at low cost. Schematically, production of "green hydrogen" is relatively simple, as shown in figure 7.2:

Green hydrogen is currently more costly to produce than gray or blue hydrogen, as electrolyzers are expensive to manufacture and require large inputs of electricity to operate. But that is rapidly changing.

The cost of electrolyzers is following a learning curve as demand and production expand, estimated at 9–13 percent.[1] This means the cost of electrolyzers is forecast to fall by approximately 11 percent for every doubling

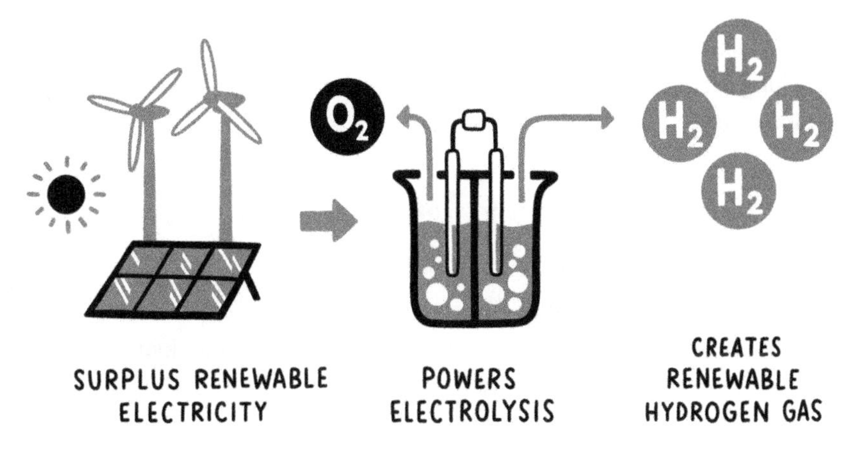

FIGURE 7.2. The production process for green hydrogen (Figure by the author)

in existing units produced. Demand for electrolyzers is forecast to grow rapidly, so even a modest learning curve will result in significantly better economics. Along with lower electrolyzer costs, the price of electricity, the primary input to creating green hydrogen, is also forecast to drop significantly because of the declining cost of power from wind and solar. Given forecasts for inexpensive electrolyzers and cheap renewable electricity, Goldman Sachs predicts a more than 500-fold increase in hydrogen production facilities by 2050.[2]

SO MANY APPLICATIONS

Dr. Julio Friedman of Columbia University foresees a broad range of applications for green hydrogen, calling it the "Swiss Army knife of deep decarbonization."[3] Hydrogen can be used in an extraordinary number of ways.

In transportation, fuel cells convert hydrogen to electricity, which is then used to power vehicles. Hydrogen-fueled buses are already in use, and hydrogen can be used to power long-distance trucking and shipping. Even aircraft. Airbus has announced plans to develop a commercially viable hydrogen airplane that could enter service by 2035.[4]

In agriculture, hydrogen is a feedstock to ammonia, and green hydrogen can be used to produce inexpensive, emissions-free fertilizer.[5] In industry, hydrogen can be used in the production of steel in lieu of coal, enticing investors to finance a $3 billion green hydrogen plant in Sweden to produce emissions-free steel.[6] And green hydrogen can replace natural gas in peaker plants to generate electricity when solar and wind farms are off-line, providing a long-term energy storage solution. The potential of green hydrogen is extraordinary, but so are the challenges.

CHALLENGES TO SCALING GREEN HYDROGEN

The technology for producing green hydrogen already exists, but it faces two significant hurdles: cost and infrastructure. The cost of producing green hydrogen, $3 to $8 per kilogram, is uncompetitive with gray hydrogen and other fuel sources and will remain so until costs decline to $1/kg.

Forecasts for green hydrogen production costs vary widely. Morgan Stanley predicts green hydrogen sited next to wind farms in the American Midwest could be competitive in 2022,[7] while Bloomberg New Energy Finance predicts it will take until 2050 for green hydrogen to decline in price to $1/kg.[8] There is little doubt that the cost of producing green hydrogen will decline markedly with cheaper electrolyzers and ever-cheaper wind and solar power. But then low-cost hydrogen will face a second challenge: transporting it from where it is produced to where it will be used.

Hydrogen is challenging to ship and store, requiring pressurization at low temperature. Existing natural gas and oil pipelines cannot be repurposed, as pure hydrogen creates brittleness in steel pipes and valves. Building new hydrogen infrastructure creates a Catch-22—should companies invest in hydrogen transportation infrastructure before the cost of green hydrogen becomes competitive or should they wait, in which case costs may never decline? Overcoming the infrastructure challenges facing green hydrogen requires innovative solutions by businesses and investors. One American company, Air Products & Chemicals, is betting it has found a way forward.

COMMERCIAL INNOVATION

Air Products, a global leader in industrial gases, has established a $5 billion joint venture with a Saudi renewable energy company to build the world's largest green hydrogen project. Located in the desert of northwestern Saudi Arabia,[9] the project is in an ideal spot for generating extremely low-cost solar power during the day and wind power at night. It also has a port.

Air Products is using an innovative strategy to address the infrastructure challenge of transporting hydrogen from Saudi Arabia to markets where it can be used. Instead of compressing and shipping hydrogen by pipeline, this facility will first convert the hydrogen to ammonia, which is denser and less costly to transport by ship. The shipped ammonia will be unloaded and trucked to refueling stations, where it will be disassociated to yield hydrogen, thus avoiding construction of costly pipelines on either end.[10] Refueling stations will provide hydrogen to buses and automobiles running on fuel cells. Air Products expects this single facility to produce enough green hydrogen to run 20,000 hydrogen-fueled buses.[11] Simon Moore, a vice president at Air Products, had this to say about his company's plan: "No kidding, this can be done."[12]

THE GREEN HYDROGEN FUTURE

Hydrogen could meet up to 24 percent of the world's energy needs by 2050.[13] For this to happen, massive amounts of additional renewables will need to be constructed to power electrolyzers, and new shipping, pipeline, and refueling infrastructure will need to be built, all of which will require investment capital. McKinsey & Company forecasts global investment in green hydrogen will reach $300 billion annually by 2030.[14] Investment bank Evercore estimates $2 trillion in spending on hydrogen from 2030 to 2050.[15]

Green hydrogen holds the potential to decarbonize the gaps remaining after the transition from fossil fuels to renewable wind and solar, and after the transition from the internal combustion engine to electric vehicles. Long-distance trucking, shipping, air travel, heavy industry, and agriculture can all reduce greenhouse emissions with green hydrogen. Most important, hydrogen can provide long-term energy storage to allow for 100 percent penetration of intermittent renewable power on the electrical grid.

The climate solutions described in section 2 of this book, including green hydrogen, can reduce global greenhouse gas emissions by a remarkable 75 percent by 2050.[16] Which is an extraordinary reversal after 250 years of emissions growth. But it is not enough to take emissions to *zero*, the target that scientists forecast is needed to avoid catastrophic climate change. That will require one more climate solution: carbon removal.

"We're definitely going all in. This is going to be huge for us."

FIGURE 8.1. Vicki Hollub, President and CEO, Occidental Petroleum (Source: REUTERS / Alamy Stock Photo)

Chapter Eight

CARBON REMOVAL

Avoiding catastrophic climate change will require a reduction in greenhouse gas emissions to *zero*, which is nearly impossible as some emissions will continue even with a rapid transition to low-carbon technologies. The only climate solution for these remaining emissions is to remove greenhouse gases from the atmosphere through "negative emissions" using carbon removal solutions, as shown in figure 8.2.[1]

Three carbon removal solutions are promising: carbon capture and storage technologies prevent CO_2 from entering the atmosphere; carbon sequestration uses biomass to absorb and store CO_2; and direct air capture applies advanced engineering to remove CO_2 already in the atmosphere. Each of these solutions presents unique advantages and challenges.

CARBON CAPTURE AND STORAGE

CO_2 can be captured at the point of generation and then stored underground, a process called carbon capture and storage (CCS). Current CCS technologies can capture CO_2 emissions from fossil fuel–fired power plants and other industrial sources, which is then transported by pipeline to an appropriate storage site. There are several options for storing CO_2; the most common is to inject it into deep, underground geologic formations such as former oil and gas reservoirs. More than 50 CCS facilities exist globally, of which 10 are in the United States.[2]

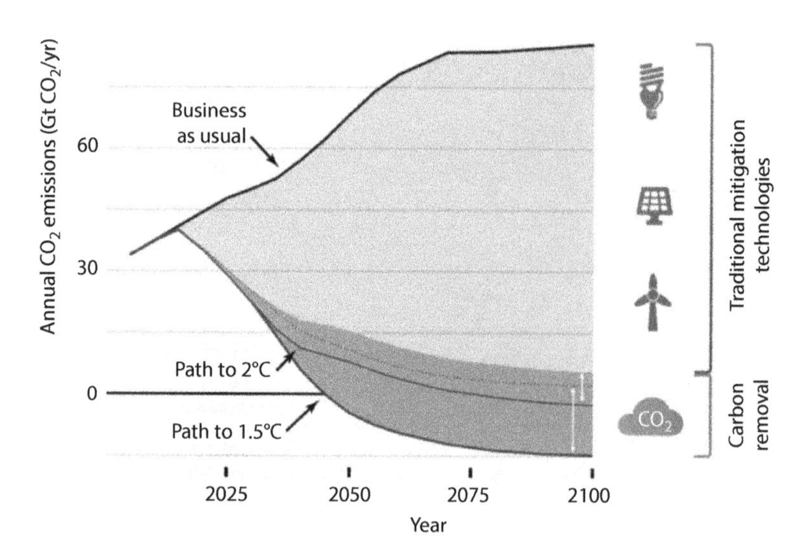

FIGURE 8.2. (Source: MCC Common Economics Blog)

The problem with CCS is that it is costly to capture, transport, and store CO_2, ranging from $60 to more than $150 per ton.[3] Costs for generating electricity from fossil fuel plants that use CCS are estimated to be 45–70 percent higher than for conventional power plants.[4] The solution to the high cost of CCS is to find a use for the captured CO_2.

CO_2 can be injected into declining oil wells, increasing oil production in a process called enhanced oil recovery (EOR). EOR sequesters CO_2 but also increases oil output, making it at best a climate change neutral activity. But EOR is an attractive commercial solution to the high cost of capturing CO_2, and 90 percent of CCS projects include EOR to generate revenue.[5] In theory, CO_2 can also be used in the production of plastics, chemicals, cement, and even as a fuel source, yet in practice these applications remain experimental.

Investment in CCS is encouraged by the U.S. government in the form of a tax credit. Section 45Q of the Internal Revenue Code awards a tax credit of $35/ton for CO_2 that is captured and reused, for example through enhanced oil recovery, and $50/ton for geologic storage.[6] Encouraged by the projected demand for carbon capture, venture capital investors with a climate focus are financing CCS technologies, and infrastructure funds are interested in the opportunity. BlackRock's Global Energy & Power Infrastructure Fund and Valero Energy are developing an industrial-scale CCS system for up to 1,200 miles of new CO_2 pipelines and storage sites, with operations to begin in 2024.[7]

Capturing and storing greenhouse gases is a compelling solution to climate change, but it is costly. A different storage solution, carbon sequestration, offers a low-cost, proven technology.

CARBON SEQUESTRATION

Carbon is sequestered through photosynthesis, the process by which trees and other plants absorb CO_2 while converting light energy into chemical energy. Carbon captured using photosynthesis is stored in trees until they die, at which point the wood decays and the carbon returns to the atmosphere. Trees are a proven climate solution that can be implemented cheaply and quickly, providing a temporary form of carbon capture.

Alternatively, plant matter can be burned as a fuel source in a thermal generator to produce electricity, and the CO_2 emissions can be captured and stored using CCS, as described earlier. This form of carbon sequestration is referred to as BECCS: bioenergy carbon capture and storage. With BECCS, greenhouse gas emissions are permanently sequestered.

The advantages of forests as a climate solution are simplicity, low cost, and scale. Carbon can be sequestered by planting trees at scale for as little as $5/ton,[8] dramatically lower than other forms of carbon capture (as a point of comparison, CCS costs $60 to $150 per ton). Planting trees also provides many co-benefits including better water quality and reduced soil erosion.

Protecting trees from being cut down, called *avoided deforestation*, is the simplest climate solution of all. Deforestation is a significant contributor to climate change because wide swaths of forests are disappearing, primarily in tropical countries. In the Amazon basin, deforestation affects more than 4,000 square miles annually, which works out to a football field's worth of trees cut down every 15 seconds.[9] Loss of forest cover from deforestation and degradation accounts for approximately one-eighth of global greenhouse gas emissions.[10]

Protecting forests and planting trees is, in theory, a highly effective and inexpensive climate change solution. In practice, carbon sequestration with forestry is surprisingly difficult to do.

Implementation Challenges

Deforestation occurs primarily for economic reasons, as trees are cut for wood and to clear land for agriculture, so preventing deforestation can reduce opportunities for farmers to earn a living, especially in developing countries where alternative forms of employment are often limited. Even worse, widespread protection of forests can lead to a shortage of land for farming and an increase in agriculture prices, threatening food insecurity among poorer communities.

Implementing reforestation plans faces several practical challenges. Trees can burn or die prematurely, a risk called *impermanence*, and protecting a forest can simply encourage deforestation elsewhere, a risk called *leakage*. Then there is the challenge in remote regions of physically protecting large forests from loggers and small farmers. But the greatest challenge to implementation of forest sequestration is *additionality*—determining whether a forest is truly at risk of being cut down.

Investing in Carbon Sequestration

The United Nations has concluded that forest sequestration is an important climate solution: "Forests are a major, requisite front of action in the global fight against catastrophic climate change—thanks to their unparalleled capacity to absorb and store carbon. Stopping deforestation and restoring damaged forests could provide up to 30 percent of the climate solution."[11]

The implementation challenges of carbon sequestration projects are daunting, but the attractive economics and scalability of forest preservation make it an interesting if risky investment opportunity. Breakthrough Energy Ventures, a venture capital firm focused on climate solutions, has financed a carbon sequestration monitoring company because planting trees "is one of the most attractive carbon removal options ready today at scale."[12]

The opportunities for investing in carbon sequestration, and the challenges to successfully doing so, are discussed in more detail in chapter 17.

DIRECT AIR CAPTURE

Direct air capture, or DAC, uses technology to remove CO_2 from the atmosphere. A significant advantage of DAC is that projects can be sited anyplace on the planet, as a reduction in CO_2 from anywhere contributes to addressing climate change everywhere. Direct air capture can even remove the emissions of previous generations from the atmosphere.

DAC technology works by pulling in atmospheric air and running it through a chemical solution to extract the CO_2. Captured CO_2 is compressed and stored in geologic formations or used in industrial processes, as done with carbon capture and storage (CCS). The technical feasibility of DAC has been proven; the key issue is the cost of these systems at scale.[13]

DAC is a costly climate solution simply because of the extraordinary amount of energy required to run large volumes of atmospheric air across chemical compounds, in addition to the cost of chemical solutions and storage of the captured CO_2. McKinsey estimates DAC costs more than $500 per ton of CO_2 sequestered, approximately 100 times the cost of planting trees.[14] Carbon Engineering, a startup leader in DAC, believes that its technology can bring costs down to between $100 and $250 per ton.[15] That would still make it a very expensive negative emissions solution, but the flexibility of DAC makes it appealing, especially to companies with few options for reducing emissions. Oil companies are, unsurprisingly, very interested.

Occidental Petroleum and Carbon Engineering established a joint venture to construct the world's first large-scale DAC plant, designed to capture 1 million tons of CO_2 annually.[16] The plant is expected to be operational in 2023. A paper published by David Keith, a professor at Harvard University and Carbon Engineering's founder, predicts a cost as low as $94/ton. If this cost were passed on to consumers, it would add 84 cents per gallon to the price of gasoline in exchange for carbon-neutral fuel.[17]

DAC technology is a negative emissions climate solution with exciting potential and enormous challenges. Vicki Hollub, president and CEO of Occidental Petroleum, summed up DAC: "It will work. We need to just get it to scale." Getting DAC to scale will require a rapid decline in cost for a nascent technology, a risky proposition for investors. But the opportunity to create negative emissions, anywhere at any time, has begun attracting venture capital investors, described in chapter 18.

The Future of Carbon Removal

Critics of carbon removal technologies believe that it encourages further use of fossil fuels and delays the implementation of zero-emissions technologies such as renewable energy. The reality, however, is more prosaic—carbon removal will be part of the climate solution simply because there is no other realistic path for reaching net-zero emissions and avoiding catastrophic climate change.

Scientists estimate that 3–7 billion tons of CO_2 must be removed annually by 2050 to offset greenhouse gas emissions that cannot be mitigated using other technologies. To keep the global rise in temperature to less than 1.5°C–2°C, it will be critical to develop and deploy carbon removal technologies at scale.[18]

Given the need for negative emissions, Elon Musk is offering $100 million in prize money for the best carbon removal technology.[19] The XPrize Foundation, which manages the prize, has determined that "any carbon negative solution is eligible: nature-based, direct air capture, oceans, mineralization, or anything else that sequesters CO_2 permanently."[20]

Venture capital investors recognize this opportunity and have begun providing capital for carbon removal technologies despite the many challenges and risks, while more risk-averse investors are financing projects to protect forests and plant trees. And an increasing number of American companies are planning on using carbon removal technologies as part of their net-zero emissions pledges. Section 4 of this book examines investment in forestry projects, and section 5 explains net-zero pledges, the strategies used by companies to meet them, and the implications for shareholders.

Climate solutions for reducing and removing CO_2 emissions—renewable energy, electric vehicles, energy storage, green hydrogen, and carbon removal—offer a path for avoiding catastrophic climate change. Even better, these climate solutions are connected in a synergistic way that will accelerate their growth and implementation. Understanding those connections is the topic of the next and final chapter in section 2.

"The Stone Age did not end for lack of stone, and the Oil Age will end long before the world runs out of oil."

FIGURE 9.1. Sheikh Yamani, Minister of Oil for Saudi Arabia (Photographer: David Levenson, courtesy of Getty Images)

Chapter Nine

BETTER TOGETHER

Sheikh Yamani, architect of OPEC during the energy crisis of the 1970s, famously predicted that oil would be replaced by better products, just as oil had previously replaced inferior sources of energy. Yamani passed away in 2021 at the age of 90.[1] He lived long enough to witness the changes he had predicted—the year before he died, integrated oil supermajor BP announced that global oil demand had peaked and could fall by as much as 50 percent over the next twenty years.[2] But even the prescient Yamani did not predict how the end of the Oil Age would represent only one of many new climate solutions, each of which developed for unique reasons, yet all of which are connected in a synergistic way. Understanding how the climate solutions are interconnected is key to understanding investing in the age of climate change.

CONNECTING THE CLIMATE SOLUTIONS

Generation of low-cost electricity from wind and solar is increasingly paired with low-cost lithium-ion batteries to provide inexpensive and reliable power to consumers and businesses. Those same batteries form the core of electric vehicles (EVs), transitioning the automobile sector from the internal combustion engine to EVs. And those EVs can provide backup power to the grid to reduce the risk of intermittency from renewables. But that's only one-half of the story.

The growth in electric vehicles means, of course, that there is growing demand for electricity. A study at the University of Texas found that the electrical grid will need 25 percent more power generation if all Americans switch to EVs.[3] That additional demand for electricity will be inexpensively provided by solar and wind. More solar and wind power means lower prices, as the learning curve drives down the cost of renewable power generation, and lower prices for electricity means higher demand for electric vehicles, which become even less costly to refuel.

Electricity is also the primary input for production of green hydrogen, creating a fuel for sectors of the economy that cannot be electrified. Low-cost renewables are the key to production of green hydrogen, and this will require rapid growth in renewables to power the electrolyzers to meet demand. Bloomberg estimates that growth in green hydrogen could increase the demand for wind and solar 30-fold by 2050, further reducing the cost of renewables.[4]

Finally, where mitigation cannot be accomplished by use of other technologies, low-cost electricity will power the carbon capture technologies necessary for negative emissions. Carbon capture and storage (CCS) and direct air capture (DAC) projects require massive inputs of power to remove CO_2 from the atmosphere. Studies estimate that the demand for DAC projects to avoid catastrophic climate change will use 25 percent of global electricity generation in 2100.[5] The increased demand for electricity will support the construction of even more wind and solar projects, contributing to further cost declines.

These five climate solutions are connected. More important, they are connected in a virtuous cycle in which growth in any one of the solutions promotes progress in the other solutions by pushing up demand and driving down costs. Figure 9.2 illustrates the connections among the climate solutions.

Note that there are hundreds of climate solutions beyond those discussed, many of which can materially reduce greenhouse gas emissions. Project Drawdown describes how reductions in food waste, improvements in recycling, and greater access to public transit will all contribute to addressing climate change.[6] Those measures matter. But this book focuses only on sectors that are commercially investable and have the potential for a very large reduction in greenhouse gas emissions.

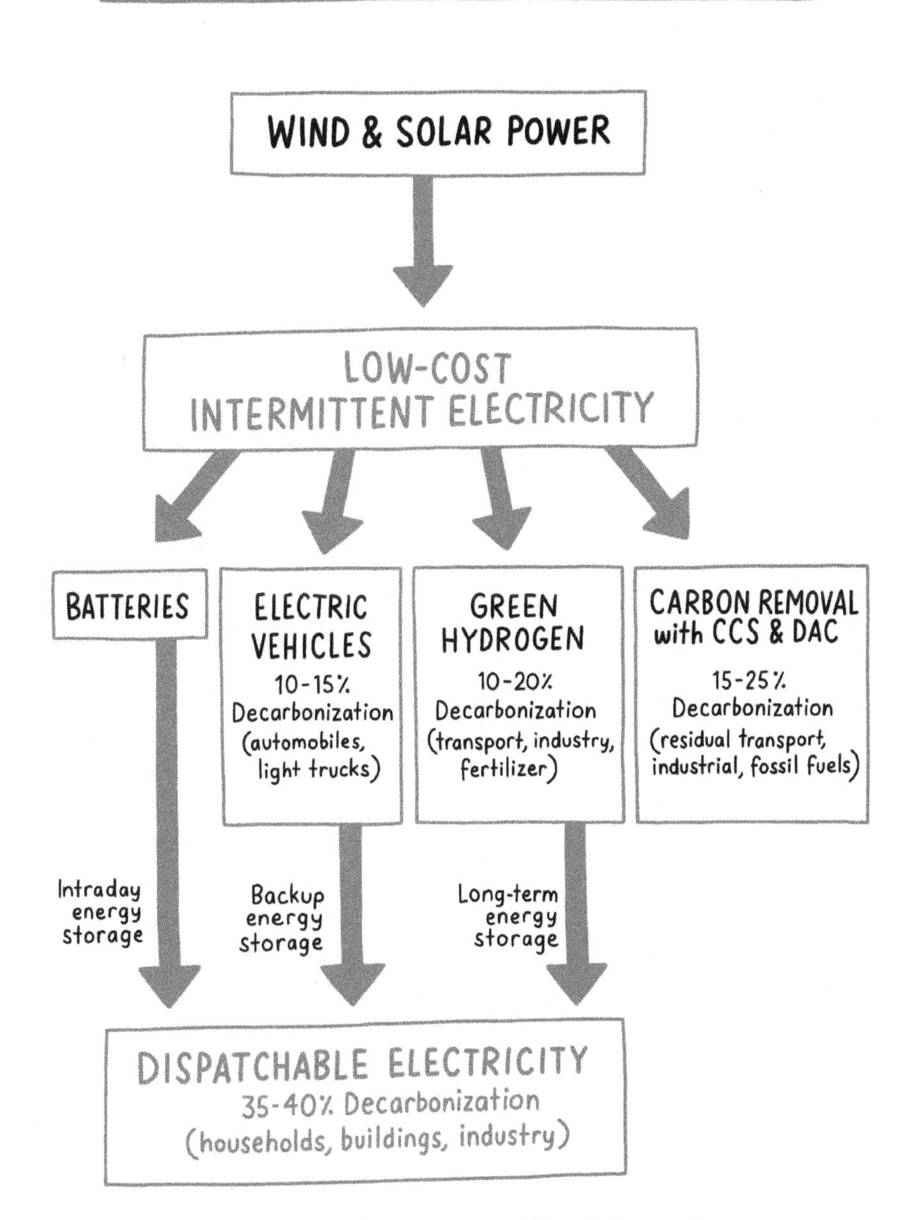

FIGURE 9.2. Better together: Climate solutions are connected (Figure by the author)

IMPLICATIONS FOR CLIMATE CHANGE

More than a century ago, Thomas Edison predicted "we will make electricity so cheap that only the rich will burn candles."[7] No one could have known just how important his prediction would turn out to be. Thanks to the inventions of Edison and so many others, humanity is on the threshold of an energy transition from fossil fuels to renewables that will electrify much of the modern economy and create the best, and likely only, opportunity to avoid catastrophic climate change.

Electricity generated by wind and solar produces zero greenhouse gas emissions. The International Energy Agency projects that converting power generation from fossil fuels to renewables and improving energy efficiency can reduce global CO_2 emissions by 38 percent by 2050.[8] Electric vehicles can reduce global CO_2 emissions by 22 percent,[9] and industry can reduce greenhouse gas emissions a further 20 percent using renewable power, energy efficiency, and green hydrogen.[10]

Green hydrogen can decarbonize sectors of the economy that are costly or impossible to electrify. Hydrogen can be used for high-temperature heat required in industrial manufacturing and in the production of chemicals, iron, and steel. Heavy trucks, ships, and airplanes cannot easily be electrified because of the weight of batteries and the long distances traveled, but hydrogen can replace diesel fuel in trucks and jet fuel in planes.

Agriculture is responsible for the remaining 10 percent of global greenhouse gas emissions.[11] Emissions from fertilizer can be eliminated using green hydrogen to produce ammonia, while land-use emissions can be abated through an end to deforestation.

Taken together, electrification and green hydrogen can reduce global emissions by approximately 75 percent. But that's not enough to avoid catastrophic climate change. Carbon removal—including carbon sequestration, carbon capture of industrial CO_2, and direct air capture—will be needed to address the remaining 25 percent to bring net greenhouse gas emissions to zero by 2050.

IMPLICATIONS FOR INVESTORS

The path to net-zero greenhouse gas emissions demands an all-of-the-above strategy to avoid catastrophic climate change, and implementing all those climate solutions requires capital, a great deal of it. Goldman Sachs estimates annual investments of $4.8 trillion through 2050.[12] By sector, the greatest demand for capital comes from the installation of renewable wind and solar power, but every one of these climate solutions is capital intensive.

The remainder of this book explains the investment strategies used by investors in the era of climate change and the specific opportunities and risks of investing capital in real and financial assets at the forefront of climate solutions.

SECTION 3

Investing Strategies

The most significant shift in investing since the Industrial Revolution has begun, moving capital away from assets that contribute to or are at risk from climate change and into assets that mitigate emissions of greenhouse gases. Section 3 explains investing strategies to redeploy capital in the era of climate change.

FIGURE 10.1. Flooding caused by Hurricane Katrina in New Orleans, the worst insured loss event in the history of insurance. Photograph by Bob McMillan of FEMA (Source: Wikimedia Commons)

Chapter Ten

RISK MITIGATION

Climate change represents a significant risk to many investments, from sudden losses due to storms to prolonged damage due to rising seas. Annual losses from extreme weather events have tripled from $50 billion in the 1980s to $150 billion in recent years, after accounting for inflation.[1] The property insurance industry is aware of climate-related risks, but until recently most investors believed climate change was too far in the future to affect the value of investments today. That is no longer the case.

TRAGEDY OF THE HORIZON

In 2015, Mark Carney, then governor of the Bank of England, made a speech to a room full of insurance executives at Lloyds of London. The speech was, as one might imagine, rather dry. But the contents were illuminating and marked a turning point in how financial firms and investors view climate risk.

Carney pointed out that climate risks were not a focus of the financial sector, as the risks are perceived to be very long-term, beyond the concern of most actors. Specifically, beyond the business cycle (quarterly to a few years), beyond the political cycle (a few years until the next election), and even beyond the purview of regulators such as central banks (two years for monetary policy, potentially a decade for the full credit cycle). Carney called this lack of concern over climate risk the "Tragedy of the Horizon."[2]

The *tragedy of the horizon* is a play on the economic term *tragedy of the commons*. As explained in section 1 of this book, the tragedy of the commons describes a situation in which a shared resource is depleted when individuals act in their own self-interest. Everyone knows a problem exists, yet no one is incentivized to address it. Carney described climate change as a tragedy of the horizon as it "imposes a cost on future generations that the current generation has no direct incentive to fix."[3]

Carney identified two critical risks that investors need to be aware of, physical risks and transition risks, and urged investors to avoid the tragedy of the horizon by addressing climate change today: "While there is still time to act, the window of opportunity is finite and shrinking."[4]

PHYSICAL RISKS

Climate change will impose physical changes on the planet, most notably a gradual rise in temperature and sea level. Global *average* temperature is projected to rise another 1°C (or 1.8°F) by 2050, but averages are misleading.[5] Carney's insight is that relatively minor physical changes in the future are creating large risks to asset values today. Evaluating this risk begins with an understanding of how a small change in the average can result in a much greater change in the probability of an extreme event.

Understanding Tail Risks

Across the United States, just 29 cities experience a month or more each year when the heat index surpasses 100°F.[6] By mid-century, scientists predict more than 250 American cities will experience such extreme temperatures, exposing millions of Americans to potentially deadly heat. Ambient temperatures above the human body's normal of 98.6°F limit outdoor activities and create serious health risks. Productivity will decline, and health care costs will go up. How can a barely noticeable change of one or two degrees of warming lead to such a significant impact?

A small shift in average temperature dramatically increases the number of days with extreme temperatures because weather patterns follow a normal probability distribution. When the average shifts, the probability of what was previously an extreme event becomes much more frequent. Figure 10.2 illustrates how a small shift in average temperature greatly increases the probability of hot weather.

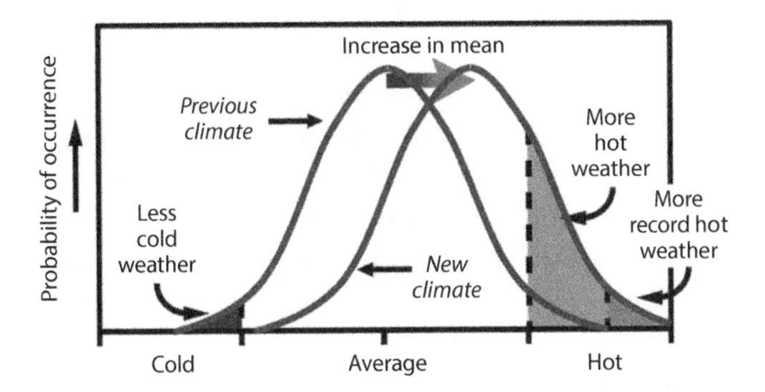

FIGURE 10.2. Normal distribution of temperature. "Climate Change 2007: Working Group I: The Physical Science Basis," IPCC, https://archive.ipcc.ch/publications_and_data/ar4/wg1/en/box-ts-5-figure-1.html;

As the normal distribution of temperatures shifts to the right, the area under the "tail" to the right increases dramatically. If "hot weather" is a day over 90 degrees, then a shift of just 1 degree in the mean temperature will significantly increase the probability of a 90-degree day. For extreme "record hot weather," the increase in probability is even greater. For example, Washington, D.C., currently experiences 7 days per year with the heat index over 100°F. By 2050, scientists predict 41 days with such extreme heat in the nation's capital.[7]

Tail risks are not unique to temperature; the same holds true for sea-level rise and flooding. Sea level is forecast to rise 6–18 inches by 2050, a modest change.[8] But flooding in some cities will increase dramatically because flooding is a tail risk. When the mean sea level increases, the tail expands, and the probability of a flooding event becomes much greater. For example, since 1960, sea level has risen only 6 inches, yet Wilmington, North Carolina, has experienced an exponential increase in flooding during that same time period, from just 1 day to 43 days per year.[9]

Tail Risk in Asset Prices

Mark Carney warned investors that "the catastrophic norms of the future can be seen in the tail risks of today."[10] To put that another way, while the physical risks from climate change will mostly occur decades in the future, asset prices are starting to reflect the increase in risk that comes from events with tail risks. In Florida, for example, tidal flooding is projected to increase from a few days to 200 days per year by 2050. Already, properties with exposure to increased flooding have lost 11 percent of their value, a total of $5 billion in losses to date.[11] That's just a hint of what is coming.

McKinsey & Company warns of a vicious cycle: insurers refuse to provide coverage to at-risk homes, banks stop offering 30-year mortgages, homes adjacent to flooded properties decline in value, and local tax bases erode. McKinsey concluded that the outcome for homeowners and investors with exposure could be grim: "Properties may see resale prices drop significantly due to severe and frequent flooding, even falling to zero if there are no prospective buyers. . . . Once buyers become aware of and price in expectations of future hazards, home prices may adjust *in advance* of significant climate-induced property destruction or flooding-related inconvenience."[12]

In other words, while the physical effects of climate change are decades in the future, the value of at-risk assets will decline well before then.

To make matters worse, asset owners will struggle to manage or hedge climate risks. Homeowners can, for now, obtain insurance to cover direct damage from flooding, but there is no insurance against falling home values. And investors in climate-exposed real estate will suffer losses along with homeowners. Real estate data provider Zillow calculates more than $200 billion worth of homes in the United States are likely to be chronically inundated by flooding by 2050.[13]

TRANSITION RISKS

Physical climate risks such as wildfires, floods, and deadly storms receive most of the media's attention, but transition risks—the potential losses resulting from new government regulations, legal liabilities, and technological change—are of greater concern to investors. Commercial banks and other public-facing investors are also exposed to reputational risk if customers find them lagging in response to climate change. But regulatory changes are the greatest transition risk, as regulations can change investment returns overnight.

As the realities of climate change become clear, it is inevitable that governments will be forced to act. And as governments have been slow to respond, when they finally do so their responses will be abrupt and disorderly because of the delay.[14] The whipsaw in policy initiatives in the United States since the 2020 election are evidence of this.

The Trump administration refused to address climate change and rolled back more than 100 environmental rules, weakening limits on greenhouse gas emissions and standards on vehicle fuel efficiency.[15] But the climate trends outlined in section 1 accelerated during Trump's presidency, and with a new government has come the inevitable policy response. President Biden's administration reversed direction, rapidly issuing executive orders requiring the Securities and Exchange Commission (SEC) and other government agencies to consider mandated reporting on climate-related financial risks and oversight of the financial system's resilience to climate change.[16]

After many years of uncertainty, most business leaders and investors welcome regulatory clarity on climate change and climate-related financial risks. The key challenge for investors: how to measure and manage climate risks?

MEASURING CLIMATE RISK

Following on Mark Carney's speech to insurance executives in 2015, regulators and investors established the Task Force on Climate-Related Financial Disclosures (TCFD) to create methodologies for financial institutions to evaluate climate risks. Chaired by Michael Bloomberg, the TCFD provides guidance on measuring climate transition risks and physical risks.

The TCFD was designed to measure the risks that climate change poses to financial stability. Companies following the TCFD guidelines evaluate transition risks (including regulatory, technology, market, and reputation risks) and physical risks (both acute and chronic risks), using scenario analysis to assess the impact on balance sheets and cash flows.[17] Scenario analysis is well understood in the financial sector, but the TCFD was the first to apply it to climate-related risks.

Measuring climate risk is made significantly more challenging by the wide range of potential paths for addressing climate change. For example, a path that transitions the global economy from fossil fuels to renewables will diminish the value of oil companies. Alternatively, a transition path that uses carbon capture to sequester CO_2 will allow for ongoing use of fossil fuels, supporting oil company operations for decades to come. From a risk measurement perspective, investors need to assess both possibilities, along with hundreds of other pathways to a low-carbon future.

Despite these challenges, investor support for the TCFD has grown dramatically. More than 600 investors managing over $65 trillion in assets have joined Climate Action 100+ to encourage the world's largest companies to report on climate risks in line with the TCFD recommendations.[18] However, the vast majority of public companies have yet to report on the financial impact of climate change on their businesses,[19] and for those companies that do report, climate risk data are most often provided in a sustainability report, not in financial filings or annual reports. This highlights the key problem for investors measuring climate risk: lack of comparable data.

In the United States, companies are not yet legally obligated to provide emissions or climate data in their financial reports, and there are no official reporting standards. This has led to underreporting, greenwashing, and difficulty comparing data across companies. A review of the TCFD in 2020 concluded "companies still aren't providing the kind of information that can reliably inform decisions by analysts and managers."[20] With this problem in mind, the IFRS Foundation, which sets international financial reporting standards, announced in late 2021 an intention to develop disclosure standards for addressing climate change.[21] Similarly, the SEC in 2022 proposed mandatory reporting of greenhouse gas emissions and disclosure of climate related risks in registration statements and annual reports.[22]

Investors will struggle with accurate measurement until regulators establish and enforce reporting standards. Nevertheless, many investors have begun to evaluate climate risk data and integrate it into their decision-making, with the objective of managing climate risk.

CLIMATE RISK IS INVESTMENT RISK

A growing number of investors agree with BlackRock's Larry Fink that "climate risk is investment risk" and are managing assets with that in mind.[23] Practically speaking, this means investors are managing climate risk using a combination of investing strategies, including selling assets facing the greatest risk, deploying capital toward companies that mitigate risk, and investing in businesses with climate solutions.[24] Institutional investors are increasingly aware of and concerned about climate risks, but there is no single agreed-upon strategy for mitigating them.

An investment strategy focused on managing climate risk is, by default, a defensive investing strategy. But defensive does not mean below-market returns. Asset manager AllianceBernstein partnered with Columbia University's Climate School to connect climate risks to financial statements and to integrate climate risks within investment analysis.[25] Portfolio Manager Michelle Dunstan explains the rationale: "Companies that are thinking robustly about addressing climate change are also minimizing the risk to their cash flows—they are generally better companies, have more sustainable cash flows and will often produce better financial outcomes for our clients."[26]

AllianceBernstein is a leader in training its analysts and portfolio managers to consider climate risks, yet it is not alone. Risk mitigation is a strategy that is rapidly gaining acceptance among investors in the era of climate change.

FROM RISK TO OPPORTUNITY

Peter Diamandis, founder of the XPrize and Singularity University, is fond of saying "the world's biggest problems are also the world's biggest business opportunities."[27] Climate change poses many risks to investors, but it also offers the opportunity for much gain—not to mention the opportunity to help solve the greatest problem of the twenty-first century. The remainder of section 3 explains investment strategies in the era of climate change that are designed not only to minimize risks but also to maximize gains.

"The logic of divestment couldn't be simpler: if it's wrong to wreck the climate, it's wrong to profit from that wreckage."

FIGURE 11.1. Bill McKibben (Source: Wikimedia Commons)

Chapter Eleven

DIVESTMENT

Refusing to invest in fossil fuel companies—divestment—is a controversial but increasingly popular climate change investment strategy. Divestment is an intuitively appealing strategy for many, as it aligns the personal values of investors with their actions. Historically, divestment was done purely for ethical reasons, but in the era of climate change, investors may find that divestment also leads to higher investment returns.

WHEN THE MOVEMENT BEGAN . . .

The first known proponent of a divestment strategy was a religious group, the Quakers, who observed an incompatibility between the institution of slavery and their religion. Quakers professed to believe that "the light of God resided in all men and women,"[1] and slavery stood against their ideals of equality.[2] In 1776, the Society of Friends in Philadelphia made a formal decision that participation in the slave trade would justify expulsion from their community, becoming the first organization to take a stand not only against slavery but also against profiting from the slave trade.[3] A divestment strategy allowed the Quakers to remain true to their ethical values and religious convictions, while providing moral support to abolitionists fighting to end the practice of slavery in America.

The modern divestment movement began in the 1970s when college students opposed to racial segregation petitioned university endowments to sell shares in companies doing business in South Africa, an apartheid regime.[4] Hampshire College became the first to commit to divesting from companies with commercial ties to South Africa,[5] and by 1988, over one hundred universities and colleges had divested, with the University of California alone selling $3.1 billion worth of public company shares.[6] F. W. de Klerk, the last president of South Africa under the apartheid regime, credited divestment with having brought about change, revealing "when the divestment movement began, I knew that apartheid had to end."[7]

DIVESTMENT AND CLIMATE CHANGE

In 2008, author and environmentalist Bill McKibben founded 350.org, a nonprofit dedicated to ending the use of fossil fuels. As part of this effort, 350.org launched a campaign to encourage college and university endowments to divest from investments in the fossil fuel industry as a strategy to defund polluting companies and address climate change.[8] By 2013, hundreds of college campuses across the United States had active divestment movements.[9]

The objective of the fossil fuel divestment movement was to replicate the success of the anti-apartheid campaign against South Africa by reducing the ability of targeted companies to finance their operations. And like that campaign, the rationale made for divestment was a moral one. McKibben explained the logic of divestment in stark terms: "If it's wrong to wreck the climate, it's wrong to profit from that wreckage."[10]

Climate change is a uniquely troubling moral issue because the individuals responsible for greenhouse gas emissions today are unlikely to suffer the damage their pollution is causing. In economic terms, this makes the impact of these emissions a negative externality, and in the case of climate change the externality is especially challenging because the impact of climate change falls disproportionately on people who have not contributed to the problem: developing countries and future generations.

Greenhouse gases are emitted mostly by wealthier, developed countries. The United States and Europe, with only 11 percent of the global population, account for 47 percent of cumulative CO_2 emissions, while the worst impacts of climate change are forecast to affect the much greater number of people living in developing countries.[11] Perhaps even more troubling, contributions to climate change effectively steal from future generations, as the impacts will primarily be felt several decades hence.

In 2014, Stanford became the first high-profile university to divest from fossil fuel investments, announcing that the endowment would no longer invest directly in coal-mining companies.[12] Stanford's president wrote that the university's review had determined "coal is one of the most carbon-intensive methods of energy generation and that other sources can be readily substituted for it. . . . Moving away from coal in the investment context is a small but constructive step."[13] But Stanford refused to divest from fossil fuel companies beyond the coal sector, worried that doing so would reduce the returns on the endowment portfolio.[14]

ARGUMENTS AGAINST DIVESTMENT

While supporters of fossil fuel divestment claimed a moral imperative, opponents of divestment argued that it would be both costly and ineffective.

Modern portfolio theory, the foundation of asset management, posits that diversification across all potential assets can improve risk-adjusted returns as long as asset returns are not perfectly correlated. Put simply, holding assets in many sectors is desirable for investors who wish to maximize their risk-adjusted investment returns. Therefore, artificially limiting the investable universe, for example by divestment, reduces diversification and could lower risk-adjusted returns.

Research published in 2015 found that a fossil fuel divestment strategy reduced the returns of investment portfolios by approximately 0.5 percent per year.[15] Another study, this time of university endowments, concluded that divestment reduced returns by approximately 0.23 percent per year.[16] Both studies appeared to validate modern portfolio theory predictions and the belief that limiting the pool of potential investments would, over the long term, reduce investment returns.

Furthermore, opponents argued that fossil fuel divestment would be ineffective at reducing access to financing by fossil fuel companies—the original objective of 350.org—or reducing emissions of greenhouse gases. For every investor divesting shares, another investor is purchasing them, leaving no impact on the companies themselves. In theory, if enough investors were to refuse to purchase fossil fuel company shares, the value of those shares would decline, and fossil fuel companies might find it difficult to raise additional capital. But many fossil fuel companies are profitable, and even those that require additional funding have found willing investors with a different moral perspective on climate change.

Opponents of divestment also argued that ongoing engagement between investors and fossil fuel companies could produce better outcomes. By divesting, investors lose the right to vote shares at the annual general meeting of company shareholders, and the management of a fossil fuel company is more likely to respond to requests from current shareholders than from investors who have divested their shares. Legendary endowment manager David Swensen, known for overseeing consistent outperformance of Yale's $30 billion endowment,[17] put it this way: "Direct dialogue with its managers is the most effective means of addressing climate change risk in the portfolio."[18]

The arguments against divestment were effective, for a while. Among the largest endowment funds, Harvard University, the University of California System, and Princeton University all rejected divestment, as did many other smaller colleges.[19] Of the few endowment funds that divested, most did so only partially, divesting coal companies while continuing to invest in the much larger oil and gas sector. The moral argument made by proponents of divestment had stirred students on college campuses but had mostly fallen on deaf ears when it came to the investment managers of those very same college endowments.

STRANDED ASSETS

Traditional fund managers feared that a divestment strategy would reduce their financial returns. However, with a focus only on returns, those investors may have overlooked the risk that asset values will decline if climate trends—expansion of government regulation over greenhouse gas emissions, rapid innovation in low-carbon technologies, changes in the physical environment, and evolving social norms—come to pass. One of those risks is the concept of "stranded assets."

Oxford University has examined the risk of stranded assets to investors, which would occur if fossil fuel companies could no longer sell their reserves. Reserves represent the amount of coal, oil, and natural gas that companies expect to profitably extract and sell in the future. Investors have traditionally viewed reserves as a valuable asset of fossil fuel companies, assigning higher share values to companies reporting larger reserves.[20] But investors may have miscalculated.

Total reserves currently held by fossil fuel companies are estimated to contain just under 3,000 gigatons of potential CO_2 emissions.[21] However, if all those reserves are burned, the resulting emissions will cause a catastrophic warming of the planet. Researchers have concluded that a third of oil reserves, half of gas reserves, and more than 80 percent of current coal reserves must remain in the ground to meet a target of 2°C.[22] As discussed in section 1, changes in consumption patterns and potential changes to government regulation are likely to render many of those reserves uneconomical, potentially stranding "unburnable" assets. As of 2020, analysts estimate that oil and gas companies hold up to $900 billion in stranded assets, representing one-third of their market capitalization.[23] Researchers at Stanford University concluded that investors own mispriced assets.[24]

REASSESSING DIVESTMENT

Investor interest in fossil fuel divestment and the risk of stranded assets brought additional scrutiny to research on this investment strategy. Investors noticed that the two early studies on the cost of divestment had both been commissioned and financed by the Independent Petroleum Association of America, calling into question the objectivity of those early reports.[25] Investors began to reassess the divestment strategy.

Further research found that financial returns are not systematically different for portfolios divested from fossil fuel stocks. A 2017 peer-reviewed study reported that the earlier studies "cherry-picked historical subperiods and indices" and that "fossil fuel divestment has not significantly impaired financial performance of investment portfolios."[26] The researchers discovered that fossil fuel stocks "only provide relatively limited diversification benefits . . . fossil fuel stocks are more or less substitutes for the market index (which has a beta of 1)"; in other words, they provide limited benefit under modern portfolio theory.

Outside of academia, famed British investor Jeremy Grantham, founder of fund management company GMO, gave a talk in 2018 titled "The Mythical Peril of Divesting from Fossil Fuels." Grantham reported that annualized returns of the S&P 500 including and excluding energy stocks were not materially different, ranging from −7 bp to +3 bp in the period between 1925 and 2017.[27]

EMBRACING DIVESTMENT

After a slow start, divestment of fossil fuel companies has become increasingly popular among college and university endowments. Moreover, divestment commitments have expanded beyond a narrow focus on coal companies to include the much larger oil and gas sector. In 2019, the entire University of

California System announced full divestment,[28] including indirect investments, stating "as long-term investors, we believe the university and its stakeholders are much better served by investing in promising opportunities in the alternative energy field rather than gambling on oil and gas." Brown University announced comprehensive divestment in 2020, stating "investments in fossil fuels carry too much long-term financial risk."[29] In early 2021, Columbia University went beyond divestment of thermal coal to include oil and gas companies, announcing it may make exceptions for companies with "credible plans for transitioning their businesses to net zero emissions by 2050."[30] And in September 2021, after years of rejecting student and alumni petitions for divestment, Harvard University announced it would no longer invest in fossil fuel companies.[31]

University and college endowments have provided early leadership on divestment, but their investment funds, at $600 billion for all U.S. college and university endowments,[32] pale in comparison with the $23 trillion in capital managed by U.S. pension funds.[33] Predictably, pension funds have also begun to embrace divestment. Concerned about the long-term risks of climate change on portfolio value, especially the risks of government regulations and advancing technologies on fossil fuel assets, pension funds are moving to the forefront with their divestment strategies. New York City Pension Funds, with $189 billion in assets, announced plans to divest in 2018,[34] followed by an announcement by New York State's $226 billion Common Retirement Fund that it will divest from fossil fuels within 5 years and from shares in other companies that contribute to climate change by 2040.[35] In total, a staggering $40 trillion in investor capital has divested from or refused to invest in fossil fuels.[36]

A BEGINNING . . .

Early opponents of divestment argued that it would be ineffective at reducing access to capital by fossil fuel companies. And for many years it appeared that the goal of 350.org to defund polluting companies was aspirational at best. That is changing. In 2020, the *Financial Times* reported coal companies were experiencing shrinking access to finance,[37] and the International Energy Agency reported difficulties financing new coal mines.[38] Investment bank Goldman Sachs warned that the financing challenges experienced by coal companies are reaching the entire fossil fuel sector.[39]

In the nineteenth century, Thomas Clarkson wrote of the abolition movement: "The greatest works must have a beginning. . . . However small the beginning and slow the progress may appear in any good work which we may undertake, we need not be discouraged as to the ultimate result of our labours."[40] The divestment movement also started slowly, taking nearly a decade to display any substantive impact. Yet it is now reducing capital flows to the fossil fuel sector. What divestment fails to do, by virtue of its strategy, is to increase capital flows toward companies focused on improving sustainability and addressing climate change. That requires a different strategy, called ESG investing.

"I propose that you, the business leaders . . . and we the United Nations initiate a Global Compact of shared values and principles, which will give a human face to the global market"

FIGURE 12.1. Kofi Annan, UN Secretary-General. (Source: Wikimedia Commons)

Chapter Twelve

ESG INVESTING

Divestment is a negative investing strategy, keeping unwanted holdings out of a portfolio. ESG is a more sophisticated investing strategy, using environmental, social, and governance factors in the investment research process to minimize risk and maximize returns. Building on the history and lessons of the divestment movement, ESG has quickly grown to encompass trillions in assets, propelled by the risk and opportunities of climate change.[1] Surprisingly, it all began with a letter written in 2004 to the CEOs of 55 major investment firms by Kofi Annan, secretary-general of the United Nations (figure 12.1).[2]

FROM A LETTER TO $100 TRILLION

Kofi Annan's letter invited financial leaders, academics, and the Swiss government to a meeting in Zurich to develop guidelines on integrating environmental, social, and governance factors in financial analysis and asset management.[3] The term *ESG* was first used in the resulting report, *Who Cares Wins,* which was published with a summary of recommendations from that initial meeting that were endorsed by Goldman Sachs, Morgan Stanley, and 18 other leading financial institutions.[4] The United Nations followed up on this initial meeting by launching the Principles for Responsible Investment (PRI), a set of six voluntary guidelines beginning with "We will incorporate ESG issues into investment analysis and decision-making processes."[5]

By 2022, nearly 4,000 asset owners and managers had signed onto the PRI, representing total assets of an extraordinary $121 trillion.[6] More than half of the world's top asset managers are signatories to the PRI.[7] The concept of ESG began with Kofi Annan's desire to give "a human face to the global market," but the extraordinary growth in this investing strategy was boosted by the recognition that ESG factors can create value for companies and for investors.

THE ESG INVESTMENT STRATEGY

Research analysts and portfolio managers consider a host of factors when deciding whether to make an investment, a process called *fundamental analysis*. At a minimum, attention is paid to financial results, management quality, product and brand value, competitors, market size, and growth opportunities. An ESG investing strategy considers additional factors that are material to the investment under consideration. Evaluation of ESG factors does not substitute for, but is always in addition to, fundamental investment analysis.

ESG factors material to a company's business depend on industry, geographic location, and firm-specific considerations. Environmental factors inevitably include greenhouse gas emissions and, depending on the company, may include other pollutants, water usage, land use, as well as the downstream impacts from using a company's products. Social factors often include considerations related to a company's employment practices and the impact on customers and stakeholders. Governance factors vary by geography, tending to focus on topics such as board composition, executive compensation, and oversight.

For many investors, climate change has become the most important category within ESG analysis. The director of sustainability research at Morningstar finds "climate has become the biggest theme in ESG investing,"[8] and the PRI lists climate change as the top concern of ESG investors.

WHAT ABOUT FIDUCIARY DUTY?

Institutional investors manage capital on behalf of their clients and are legally obligated to act in their client's best interests. This is their fiduciary duty. Some investors have questioned whether ESG investing is compatible with fiduciary duty as, by its very nature, ESG investing considers factors that benefit not only the investor but also society at large. Does consideration of ESG factors violate an institutional investor's promise to put its clients' interests first?

This question was first addressed in 2005 in the *Freshfields Report*, written by a leading UK law firm. The report concluded that "integrating ESG considerations into an investment analysis so as to more reliably predict financial performance is clearly permissible and is arguably required in all jurisdictions." But many investors, especially those based in the United States, were unconvinced by the *Freshfields Report* and hesitated to engage in ESG investing strategies, expressing concern over their duty to maximize risk-adjusted financial returns for their clients. It took another decade for this concern to be addressed.

In 2015, the U.S. Department of Labor, which provides oversight of corporate pension plans, provided long-awaited guidance on the issue of ESG investing and fiduciary duty. The department concluded that institutional investors may use an ESG investing strategy "when a fiduciary prudently concludes that such an investment is justified based solely on the economic merits of the investment."[9] However, in 2020 the Department of Labor under President Trump reversed course, issuing a rule restricting plan fiduciaries from considering ESG factors. In 2021, the Department of Labor under President Biden reversed course once again, announcing "fiduciaries may consider climate change and other environmental, social, and governance (ESG) factors when they make investment decisions"[10]

Despite regulatory instability on the issue, Robert Eccles of Harvard Business School summed up ESG investing and fiduciary duty, writing "recent legal opinions and regulatory guidelines make it clear that it is a violation of duty not to consider such factors."[11] ESG factors are intended to create value for society; they can also create value for companies and for investors.

CREATING VALUE FOR COMPANIES

ESG investors require data on companies' environmental, social, and governance performance, just as companies report financial performance. Financial reporting, however, has been in existence for nearly a century, and companies have long followed standard accounting and reporting rules. In recent years, similar ESG standards have been developed, and these continue to be refined as they gain wider use.

The Global Reporting Initiative, or GRI, created global standards for sustainability reporting and is used by 80 percent of the world's largest companies.[12] Similarly, the Sustainability Accounting Standards Board, or SASB, developed standards for companies to identify sustainability issues that matter most to their investors, much as the similarly named FASB does for financial accounting standards.[13] This creates additional data gathering and reporting work for companies. But it can pay off, as focusing on ESG factors has been shown to create significant value for companies and for their shareholders.

Companies that pay attention to material ESG factors, and manage them well, tend to outperform their peers. Research from Harvard Business School finds that "strong ESG performance has a high correlation with strong valuations, expected growth and lower costs of capital."[14] Which upon closer examination is not all that surprising, as companies that perform well on material ESG factors have three significant competitive advantages: loyal employees, customers, and investors.

Research by Vanessa Burbano of Columbia Business School found that employees were willing to work for significantly lower pay at companies they believed to be socially responsible.[15] Alternatively, for the same wages as competitors, more socially responsible firms can attract higher-performing workers. This behavior is not due to having more socially oriented employees. Rather, employees "interpret an employer's social responsibility as a signal about how the employer will treat them." Burbano's research found that information on social responsibility had the greatest effect on the highest performers. In tight labor markets, where companies must compete for the best employees, strong ESG performance can provide a material recruiting and retention advantage.

In turn, customers also favor companies viewed as sustainable. A study by Nielsen of 29,000 consumers found that 50 percent were willing to pay more for products from socially responsible companies.[16] Another Nielsen survey found that 81 percent of consumers globally feel strongly that companies should help improve the environment.[17] Other studies have reached similar conclusions, finding that consumers prefer to purchase merchandise with social or environmental attributes.[18] In highly competitive markets, companies doing well on ESG factors are better positioned to attract customers.

Finally, strong ESG performance can help companies attract investors. Research by George Serafeim of Harvard Business School examined the relationship between sustainability reporting by companies and the composition of their investors. Serafeim concluded that these firms have a "more long-term investor base" with "more dedicated and fewer transient investors."[19] Later research by the investment bank Morgan Stanley found a correlation between high ESG scores and lower costs of capital for companies in both developed and emerging markets.[20] Companies that do well on ESG factors can raise long-term investment capital more cheaply than their competitors, a significant competitive advantage.

ESG creates value for companies, allowing them to outcompete rivals because employees, customers, and investors prefer firms that rank highly on sustainability metrics. These companies are also better at managing material ESG risks, including the physical and transition risks of climate change, creating a further competitive advantage. If sustainability is a good strategy for companies, is it also a good strategy for investors?

CREATING VALUE FOR INVESTORS

Active investors who select stocks with the objective of outperforming an index are increasingly using ESG strategies to improve their performance, or "alpha." In a survey, 63 percent of market participants indicated that ESG information was material to making investment decisions.[21] More important, a Harvard study found that ESG improves risk-adjusted returns: "ESG screening adds about 0.16 percent in annual performance, on average."[22] While 0.16 percent, or 16 basis points, may not seem like much, for fund managers it can be the difference between median performance and outperformance.

Other studies have found that an ESG strategy can reduce downside risk for investors. Analysis by Morgan Stanley found reductions in risk levels for sustainable funds versus those of their more traditional peers.[23] The median downside deviation for sustainable funds was 20 percent less than what traditional funds experienced.

More recently, sustainable fund performance has been resilient through the COVID pandemic.[24] BlackRock observed that during the volatile Q1 of 2020, 94 percent of sustainable indices outperformed their parent benchmarks.[25] Academic research confirmed the results experienced by BlackRock and other investors during the pandemic: "As the stock market started to collapse, an increasingly strong association between ESG and abnormal returns prevails."[26]

Given the potential for improved risk-adjusted returns, many new ESG-focused funds have sprung up to take advantage of these opportunities. Morningstar estimates there are 3,300 funds focused specifically on ESG strategies, representing $840 billion in assets under management.[27] Active managers have been quick to exploit this opening, with many seeing the still-unsettled nature of ESG reporting and performance as sources of alpha.[28]

Passive investors, who select portfolios of stocks designed to match an index, are engaging in ESG investing strategies but in a different style from active investors. Passive investors are using indices specifically designed to select companies performing strongly on ESG metrics. There has been a proliferation of index products in recent years to serve this investor demand, matching popular market benchmarks like the S&P 500 index.

While equity managers were the first to adopt ESG metrics, fixed-income managers are also working to integrate ESG metrics into their investment processes.[29] Even esoteric fixed-income products such as collateralized loan obligations (CLOs) have begun to include some ESG integration.[30] Research has found that fixed-income ESG funds, unlike equity funds, tend to be more focused on managing downside risk than reaching for higher returns, in part because issuers with strong ESG ratings may enjoy tighter credit spreads.[31]

Even private equity investors are integrating ESG factors into their due diligence and investing process. Approaches have differed across funds, with some such as Blackstone and Carlyle working to integrate ESG across their existing portfolios, while other private equity asset managers have launched new funds dedicated to an ESG investing strategy.[32] Investors are finding that environmental, social, and governance issues are highly complex, suggesting that companies that are good at managing ESG factors are well run. In other words, strength on ESG factors may reflect the quality of a company's management team.

BEYOND RESEARCH, ACTIVISM

ESG investors committed to addressing climate change are also using activism to encourage public companies to reduce greenhouse gas emissions and improve financial returns. Shareholders are owners, with the legal rights that come with ownership, an advantage that can be used to their benefit. Shareholder activism by equity investors concerned about climate change takes two forms: engagement with company management and voting at the annual general meeting.

Shareholder engagement is the practice of institutional investors meeting with companies to discuss ESG issues that could affect long-term financial performance. BlackRock reports meeting with the management teams at more than 3,000 companies every year to signal concerns, provide feedback, and share insights on climate change and other ESG issues.[33] Public companies are usually willing to engage with institutional investors, but sometimes shareholders are forced to take a more aggressive tack: voting.

Investors have the legal right to submit a proposal requesting a vote at the annual general meeting. Companies can attempt to exclude proposals they believe are not relevant to the underlying business, but the Securities and Exchange Commission (SEC) has the final say in determining eligibility for exclusion. Even small investors can engage in activism, as any shareholder of a U.S.-listed company owning more than $25,000 in shares held over 1 year can file a shareholder proposal. Support for climate-related proposals has increased markedly, from 28 percent in 2019 to 49 percent in 2021, with the most popular resolutions calling for companies to reduce greenhouse gas emissions in line with the Paris Agreement.[34]

Investors seeking votes on their proposal at the annual general meeting will often solicit proxy votes from other investors. Proxy votes designate one investor to vote on behalf of others who cannot attend the meeting. Public companies can find themselves in a proxy fight, in which one shareholder convinces other shareholders to join them to form a large block of voting shares to pass a proposal when management recommends against it. Robert Eccles and Colin Mayer of Oxford University found that "proxy voting is playing hardball, but this is a skill activist hedge funds have honed over many years of practice."[35] It took a small, oddly named investment firm called Engine No. 1 to demonstrate how playing hardball can be useful in the era of climate change.

In 2021, Engine No. 1 led a dissident shareholder group to propose four new directors supporting low-carbon strategies to the board of ExxonMobil. Engine No. 1 rallied support from major institutional investors, including BlackRock, CalSTRS, and the New York State Common Retirement Fund. To the surprise of many, the engagement campaign was successful, as shareholders holding a majority of shares voted to add three new members to Exxon's board despite strident opposition from management. The activism strategy employed against Exxon represented the first time board directors were replaced in this manner over climate change.[36] Analysts at Bloomberg concluded that Engine No. 1's engagement strategy "is another sign that ESG issues, and climate change in particular, are now mainstream."[37]

ESG's POSITIVE FEEDBACK LOOPS

Asset owners' demand for sustainable investments forced many fund managers to sign onto the Principles for Responsible Investment, compelling analysts at those funds to incorporate ESG metrics into their decision-making process. This in turn created demand from analysts for data on public-company environmental, social, and governance factors, which drove participation by those companies in standards-setting organizations such as the Global Reporting Initiative. Companies subsequently learned that by managing ESG factors, they created a competitive advantage with their employees, customers, and investors, which created a positive feedback loop that encouraged further focus on environmental, social, and governance issues. And then investors found that an ESG investing strategy could create alpha, or outperformance.

Positive investment results from ESG investing create an important feedback loop, driving further growth, as better risk-adjusted returns motivate investor pressure on companies to collect better data and conform to ESG standards to enable the next round of outsized returns. This feedback loop is so effective that simply signing onto the Principles for Responsible Investment leads to an increase in funds flowing to asset managers of 4.3 percent per quarter.[38] This may explain why nearly 4,000 firms representing $121 trillion in assets are signatories.[39] Effectively every major asset manager now participates, including BlackRock, PIMCO, Vanguard, and State Street. Goldman Sachs summed up the outcome of this positive feedback loop: "ESG investing, once a sideline practice, has gone decisively mainstream."[40]

ESG INVESTING'S IMPACT ON CLIMATE CHANGE

UBS, the world's largest asset manager for the ultrarich, reported that ESG investments are popular because "our wealthiest clients want to know their investments are making a difference to make the world a better place."[41] The assumption in that statement is that an ESG investing strategy has positive impact. The reality is less certain.

Despite the extraordinary growth in ESG investing, this strategy still faces significant challenges. Foremost among these is a lack of consistent, comparable data reported by companies on environmental, social, and governance factors. Reporting is voluntary, and companies frequently cherry-pick the data that make them look best, a tactic sometimes referred to as "greenwashing." Shiva Rajgopal of Columbia Business School analyzed corporate ESG data and found that "climate change is a real concern and firms could do well by doing good. Identifying which firm is doing well by doing good, however, is a non-trivial exercise."[42]

The SEC is taking a heightened interest in ESG information reported by companies and marketed by fund managers. In 2021, U.S. and German regulators opened an investigation into allegations that Deutsche Bank's asset management affiliate exaggerated its use of ESG criteria, which sent reverberations throughout the ESG investing sector.[43] European regulators are cracking down on fund managers inappropriately marketing ESG, and the SEC has formed a task force to investigate potential misconduct in the United States.

Even when good data are available, analysts struggle to determine which information is material for appraising financial performance and for addressing specific environmental, social, and governance problems. Evaluating the impact of ESG investing on climate change is one of those challenges.

ESG investing is effective at raising awareness of climate change and improving disclosure by companies of their environmental impact, and more than 93 percent of the world's 250 largest companies report sustainability metrics.[44] Even better, ESG investors have encouraged companies to materially reduce their emissions of greenhouse gases. For example, total carbon emissions from companies in the S&P 500 declined 11 percent from 2009 through 2017,[45] an impressive achievement given that revenue rose 36 percent over the same period.[46] However, the decline in emissions may also be attributable to the shift from manufacturing companies to services companies in the S&P 500; a direct link between ESG investing and lower greenhouse gas emissions is difficult to prove.

"THE SINGLE LARGEST INVESTMENT OPPORTUNITY IN HISTORY"

Former U.S. vice president Al Gore is renowned for his speeches on climate change and the movie *An Inconvenient Truth*, which highlight the risks of catastrophic climate change. What is less well-known is that Vice President

Gore has generated exceptional financial returns using an ESG strategy that specifically considers climate change when selecting investments.

Gore cofounded Generation Investment Management in 2004, the first major independent asset management firm to exclusively use an ESG investing strategy. In the first decade of its existence, Generation's flagship fund beat the MSCI World Index by 5.6 percent *per year*.[47] Fund investors enjoyed a more than tripling of their money during that period, and Generation was ranked among the world's top funds. The flagship fund went on to return 17.5 percent per year through 2018 versus 8.6 percent for the MSCI World Index.[48] Gore's rationale for using an ESG investing strategy is straightforward: "improving quality of life without borrowing from the future is the single largest investment opportunity in history."[49]

ESG investing is a strategy benefiting from the systemic climate trends identified in section 1 of this book: rapid innovation in low-carbon technologies, changes in the physical environment, expansion of government regulations, and evolving social norms. Generation Investment Management was one of the first funds to demonstrate the connection between those trends and investment returns, dramatically outperforming the market and growing assets under management to $39 billion by 2022.[50] The success of Generation and other early ESG investors convinced traditional asset managers of the value to be gained from incorporating climate trends in their investment strategies.

For many investors, the appeal of an ESG investing strategy is that it can be applied to an entire portfolio. But ESG is not focused on solutions. Investors seeking a more targeted approach at the intersection of capital and climate change use a different strategy, called thematic impact investing.

"Building great companies that also help to solve some of the core challenges of our age is not an oxymoron as some would posit—in our view, these goals go hand-in-hand."

FIGURE 13.1. Nancy Pfund, DBL Partners (Source: Wikimedia Commons)

Chapter Thirteen

THEMATIC IMPACT INVESTING

Thematic impact investing is a strategy to finance businesses that address a specific environmental or social challenge, such as climate change. Unlike ESG investing, thematic impact investors primarily finance private companies. Themes are selected on the basis of expertise and expectations for generating attractive risk-adjusted financial returns. A survey by Morgan Stanley found that 72 percent of thematic impact investors are seeking to address climate change.[1]

DOING WELL BY DOING GOOD

Thematic impact investing began in the late 1990s, led by pioneering and dedicated investment professionals who believed they could generate attractive financial returns while simultaneously addressing some of the world's most challenging environmental and social problems. It was a radical idea. Most investors at the time assumed a trade-off existed between doing well and doing good, believing that a thematic impact investing strategy would result in below-market returns on investment. Academic research supported this perspective, with one study concluding that " 'doing well by doing good' is an illusion."[2] With little interest from traditional asset owners, the first thematic impact investors were financial innovators who pushed against that presumption.

SJF Ventures and DBL Partners were two of the earliest thematic impact investment funds, and both initially struggled to raise capital. SJF was founded as the Sustainable Jobs Fund in 1999 with a focus on creating employment in low-income communities, later widening its themes to include climate change, health, and education. The first fund was a mere $17 million, and SJF initially struggled to find investment opportunities that created good jobs and generated strong financial returns.

DBL Partners began similarly, focusing on investing in disadvantaged communities with the objective of expanding employment. The name DBL stood for "double bottom-line"; the first representing financial performance, and the second representing social or environmental impact. DBL's initial fund, launched in 2001, took nearly three years to raise a modest target of $75 million.[3]

A decade later, however, both SJF and DBL had raised hundreds of millions of dollars for their funds, which were oversubscribed by investors keen to participate in a thematic impact investing strategy.[4] To the surprise of many, the first thematic investors had discovered that doing well and doing good was not only feasible, but profitable.

DISCOVERING ALIGNMENT

DBL and SJF struggled with their first funds but learned two invaluable lessons: thematic investors are better aligned with the entrepreneurs they fund, and those entrepreneurs create businesses that better align with their employees and customers. With this knowledge, DBL and SJF proved that thematic impact investors can generate very attractive financial returns.

In popular or "hot" sectors, investors in private companies find themselves chasing the same entrepreneurs and offering roughly the same financing terms. Because capital itself is fungible, investors must strive to differentiate themselves in their competition for the most promising deals. For many entrepreneurs, a key differentiator among investors is the values they demonstrate with respect to issues such as sustainability and climate change.

Entrepreneurs with a social mission often prefer thematic impact investors. Brent Alderfer, cofounder and CEO of a solar company funded by SJF, recalled "we were fortunate to find an investor aligned well with our company on building a values-driven enterprise."[5] Thematic impact investors have a competitive advantage over traditional venture and private equity investors because entrepreneurs with climate solutions want financing and advice from investors who share their passion.

Having attracted the best entrepreneurs, thematic investors find that climate-focused businesses have several advantages that help them prosper. Companies with a clear social mission attract better employees and more loyal customers. For venture-stage businesses, these advantages are especially important to their success, as hiring talented employees and attracting initial customers is critical to surviving the very risky early years of start-up businesses. But these competitive advantages only exist when a company's business model is truly aligned with the target impact.

ALIGNING IMPACT AND PROFITS

DBL and SJF learned that thematic investors can generate very attractive financial returns when the environmental or social mission and the bottom line are mutually reinforcing. Climate change is an excellent fit. For example, a solar power company is perfectly aligned with its mission—the more solar panels sold, the greater the positive impact on both climate change and the company's bottom line. Similarly, an electric-vehicle company sells automobiles that reduce emissions of greenhouse gases—increasing market share of vehicle sales improves investor returns and reduces the carbon intensity of transportation, a mutually positive outcome. DBL was an early investor in Tesla, in recognition of its potential for financial and environmental returns.

Thematic impact investors such as DBL have, in the words of founder Nancy Pfund (figure 13.1), "proven out the thesis that the first and second bottom lines are not mutually exclusive but are in fact mutually reinforcing."[6] DBL's first fund generated an annual IRR to investors of 24.4 percent versus 7.7 percent for comparable vintage funds and was ranked second out of the top 25 funds managing under $250 million through 2015.[7] Instead of a trade-off between doing well and doing good, thematic impact investors demonstrated they can outperform traditional investors while simultaneously addressing climate change.

IMITATION IS THE SINCEREST FORM OF FLATTERY

The first thematic impact investors were small start-up funds like DBL and SJF. For more than a decade, those funds quietly committed capital to entrepreneurs with businesses designed to address climate change and other social and environmental challenges, proving out an investment strategy of alignment between investors, entrepreneurs, employees, and customers and generating attractive financial returns. Predictably, traditional investment funds took notice.

In 2015, Bain Capital, a leading private equity fund with $105 billion in assets under management, launched the Double Impact Fund with former Massachusetts governor Deval Patrick.[8] Bain Capital announced that the new fund "will focus on delivering attractive financial returns by investing in projects with significant, measurable social impact."[9] Bain's new fund quickly received commitments for $390 million, an unprecedented amount of capital in the sector at that time.[10] Asset owners were waking up to the potential of a thematic impact investing strategy.

Bain's entry into the sector was rapidly followed by several other traditional investment funds. TPG Capital, a private equity firm with more than $100 billion in assets under management, launched the Rise Fund, a thematic impact investing subsidiary with an ambitious fund-raising target of $2 billion. The Rise Fund not only met its initial target; by 2020, it was managing $5 billion.[11] Even legendary private equity firm KKR entered the sector, launching a $1 billion Global Impact fund that "seeks to invest in opportunities where financial performance and societal impact are intrinsically aligned."[12] Traditional funds had imitated the first thematic investors in pursuing alignment between the mission of the fund, the business models of their companies, and environmental challenges such as climate change. The world's largest and most successful investors had learned that "doing well by doing good" is achievable with the right strategy.

CHANGING THE WORLD BY CHANGING PERCEPTIONS

Thematic impact funds financed several of the most iconic entrepreneurial ventures to positively affect climate change. DBL was an early investor in electric-vehicle manufacturer Tesla and residential solar company SolarCity, leaders in their respective sectors. SJF invested in Nextracker, renewable energy developer Community Energy, and many others that have contributed to a reduction of more than 2.8 million tons of CO_2 annually.[13] But the greatest impact of thematic investors is to change the perception of risk.

Entrepreneurs and start-ups require significant investment capital to establish operations and prove their business models, after which they can become profitable and generate cash flow to sustain further growth. However, traditional investors are risk-averse and hesitant to commit capital to business sectors in which they have little experience. Investors often demand a higher return on investments in unfamiliar sectors to compensate for the perception of higher risk. This creates a problem for entrepreneurs with businesses that address climate change—inexperienced investors perceive the risks to be high and therefore demand a high return on investment, increasing the cost of capital for the company. Unfortunately, a high cost of capital makes the new business uncompetitive against incumbent companies that can access financing at low rates.

The renewable energy sector is a case in point. Investment bank Lazard publishes an annual report on the renewable energy sector, which in 2011 highlighted the reduced availability, and increased cost, of capital for wind and solar projects. Renewable energy projects are capital intensive, so a high cost of capital makes them less competitive against coal and natural gas projects. But by 2020, the perception of risk in renewables had declined so dramatically that investors were, incredibly, requiring a lower investment return than for fossil fuel projects. The International Energy Agency reported that the cost of equity for oil and gas companies had increased above 12 percent while companies investing in renewable energy could access capital below 6 percent.[14]

The sector in which risk perceptions may have changed the most is electric vehicles. In 2006, DBL and a handful of other investors made what appeared to be a very risky investment in Tesla, the first all-electric-vehicle company. And by most measures it was risky, as competing electric car brands such as Fisker struggled to avoid bankruptcy.[15] But by 2020 Tesla had produced more than 1 million cars and kickstarted the electric car industry.[16] Analysts at Bloomberg estimate there will be more than 500 different EV models globally in 2022, creating an impact far beyond what any one firm could accomplish.[17]

Thematic impact investors such as DBL and SJF have played a critical role by changing the perception of risk in renewable energy, electric vehicles, and other industries designed to mitigate climate change, driving down the cost of capital and increasing the competitiveness of those businesses. Traditional investors, seeing the commercial success of renewable energy and electric-vehicle companies, have learned that the actual risk of investment has been significantly lower than the perceived risk and are providing capital to climate-focused entrepreneurs on increasingly attractive terms. Thematic impact investors have accomplished more than simply financing some of the world's most iconic companies; they have changed the way that traditional investors fund entire sectors.

Thematic impact investors are opening up capital to an ever-increasing range of climate solutions. But some technologies and innovations are so risky, and the returns so far in the future, that even the most aggressive thematic impact investors are unable to commit. Those opportunities require an entirely different strategy, called impact first investing.

"We are willing to wait a longer time for returns than other funds."

FIGURE 14.1. Bill Gates, Founder, Microsoft; and Founder, Breakthrough Energy Ventures (Source: Wikimedia Commons)

Chapter Fourteen

IMPACT FIRST INVESTING

Impact first investors focus on solving social and environmental problems and are willing to accept a below-market financial return in exchange for greater impact. An impact first strategy is, therefore, not a better way to invest, but it might be a better way to solve climate change.

FROM PHILANTHROPY TO PHILANTHROCAPITALISM

In the 1970s, philanthropists began to discuss an unusual idea: making investments instead of grants to solve social problems.[1] The U.S. tax code was changed to allow foundations to make investments, as long as they served a charitable purpose and expected a below-market return. The Ford Foundation began to explore this new opportunity, termed *program-related investments*, or *PRIs*, as an alternative to making grants.

Foundations traditionally make grants to nonprofits to further their social mission, so the idea of investing in a nonprofit is counterintuitive, which may explain why PRIs were slow to gain acceptance. But the Ford Foundation recognized that PRIs could further their mission in two significant ways. First, if the PRI generated any financial return, even a below-market return, the dollars received back could be later reinvested or donated as a grant, multiplying impact and furthering the foundation's social mission. Second, successful stewardship of the investment could signal to traditional investors that the nonprofit is creditworthy, providing access to commercial capital.

Some donors also believed that servicing debt would bring additional rigor to the operations of nonprofits receiving PRI loans, improving the efficiency and effectiveness of nonprofit operations.

The first PRIs were loans, as nonprofits are structurally prohibited from issuing shares for equity capital. Eligible borrowers were nonprofits with income-generating models, such as affordable housing, and loans were made at zero or low interest rates, clearly below-market yet with an expectation of repayment. Although PRIs were designed for U.S. foundations, many of the initial investments were overseas.

Early experiments with PRIs were refined by organizations in developing countries making small loans to the poor as a tool to reduce poverty. Grameen Bank, in Bangladesh, was one of the first of these lenders. When the founder of Grameen Bank, Muhammad Yunus, first proposed making loans to poor women living in rural villages, bankers laughed at his request. and Yunus concluded that "the poor remain poor because they have no access to capital."[2]

Yunus proved the bankers wrong. Grameen Bank made relatively low-interest loans to the poor and grew to serve 9 million borrowers in Bangladesh, 97 percent of whom are rural women, while earning a small profit.[3] Grameen Bank demonstrated that a "low-profit" model could be both financially sustainable and enormously effective. In 2006, Yunus became the first person to win the Nobel Peace Prize for a commercial endeavor, and microfinance grew into a global industry of more than 10,000 firms providing loans to 140 million borrowers who previously had no access to capital.

The extraordinary growth and success of the microfinance sector demonstrated that financial tools could be used to address social needs, creating a shift in how philanthropists thought about investing and impact. Donors transitioned from making only grants to also providing low-interest loans, and then to investing equity capital in for-profit companies with a social mission. Making loans was challenging, but equity investments was even more complex. To meet the growing interest of philanthropists, finance professionals established impact first funds.

This model was pioneered by the work of the Acumen Fund. Started by Jacqueline Novogratz in 2001, Acumen sought to bring out the best of market-based scalability with a philanthropic focus on the world's poorest citizens. Acumen did this by accepting higher-risk investments and longer time horizons, what it termed *patient capital*.[4] It coupled those investments with intensive oversight and support for its portfolio companies to help them achieve their impact potential.

In 2006, *The Economist* pronounced "The Birth of Philanthrocapitalism," describing the ways in which donors were using traditional investment tools, loans and equity investments, to address the world's social and environmental challenges.[5] The following year, a meeting of donors led by the Rockefeller Foundation coined the term *impact investing*, defined as "investments made with the intention of generating both financial return and social or environmental impact."[6] This launched a new strategy to tackle climate change, combining the efficacy of investing with philanthropic capital.

REVISITING BELOW-MARKET RETURNS

Foundations in the United States making program-related investments, or PRIs, are required by law to earn a below-market return. Technically, the expectation of below-market return would be satisfied if a prudent investor seeking a market return would not enter into the investment. In the early years of impact first investing, this was achieved by foundations making loans at very low or zero interest rates, clearly below-market in relation to bank loans or other forms of commercial capital. Alternatively, an investment can be categorized as below-market if it is very risky. For philanthropic organizations focused on climate change, funding high-risk investments became the favored form of impact first investing.

Risk is defined as the probability that the financial return on an investment will differ from its expected return. The lower the probability, the higher the expected return must be to entice investors to commit capital. Risk comes in many forms: investing in a new sector, an unstable country or region, an unproven technology or business model, an innovative investment structure, or illiquidity. Traditional investors seeking market returns will avoid risky investments when the risk-adjusted return, that is, the expected return after consideration of all the risks, falls below returns from less risky investments. Impact first investors, on the other hand, will invest in assets that are risky; in fact, the law compels them to do so. Remarkably, those investments will occasionally generate high returns.

Acumen Fund made an impact first equity investment in d.light, founded in San Francisco in 2007 to develop low-cost solar-powered lamps to replace kerosene lamps in homes in developing countries that lack electrical grids.[7] Solar lamps are safer and less polluting than traditional kerosene lamps, reducing greenhouse gas emissions while improving customers' lives. Structured as a for-profit company with a social impact mission, d.light's business model was unproven. The company was initially financed by a combination of grants, winnings from venture competitions, and investment by Acumen.[8]

As it gained traction, d.light raised successive venture rounds from increasingly larger investors, totaling $197 million in capital over many years.[9] Once d.light proved its business model, Acumen began to exit its investment, selling a portion of its stake in 2018 for a return that was 2.4 times the invested capital.[10] Acumen then reinvested the financial gains from d.light in a fund dedicated to financing companies delivering clean energy to underserved communities.[11]

Acumen's investment in d.light was very risky, and in 2008 no commercial investor would fund the company. Nevertheless, Acumen more than doubled its money, beating the average venture capital fund return over the same time period.[12] For Acumen, d.light's success had an impact beyond a financial return, providing zero-emissions lighting to 100 million people in more than 70 countries.[13]

CROWDING-IN COMMERCIAL CAPITAL

Impact first investors who are focused on climate change seek to create change in two ways: by financing businesses that reduce emissions of greenhouse gases, and by demonstrating that new technologies, business models, and sectors are investable. The first objective, growing companies such as d.light, is important, as businesses scale much faster than most government-led or nonprofit initiatives, creating immediate impact by reducing the use of fossil fuels and greenhouse gas emissions. But the much greater impact comes with the second objective, revealing to commercial investors that these opportunities are attractive on a risk-adjusted basis.

The limitation of an impact first investing strategy is in the relatively small amounts of capital available for investment, as traditional investors will not—in fact, cannot if they are fiduciaries—commit to investments with below-market returns. Therefore, the pool of impact first capital is limited to philanthropists. In the United States, total philanthropy averages $450 billion per year, a small fraction of the $100 trillion in financial capital available.[14] Impact first investors are simply too small and underfunded to solve a massive problem such as climate change, therefore their objective is to use relatively small amounts of impact capital to catalyze or "crowd-in" commercial investors.

Crowding-in is an economic term describing the process by which development organizations inspire private investment that otherwise would not have taken place. This is most often achieved by seeding small investment funds to fill gaps in the commercial market, making investments that traditional investors believe to be too risky. If successful, these investment funds demonstrate that investors are missing an opportunity for an attractive return, thereby catalyzing commercial capital.

Attracting commercial capital to a new and unproven technology, sector, or country is never easy, often taking many years. In the example of d.light, Acumen invested for a decade before the first opportunity to earn a return and demonstrate the profitability of the investment. To accelerate the financing of climate change solutions, impact first investors employ several tactics to attract commercial capital. These tactics, referred to as *blended finance*, include the following:

- Providing first-loss capital to decrease risk to commercial investors. This is structured by placing the impact first investment lower in the capital stack; for example, as common equity instead of preferred equity.
- Offering higher returns to commercial investors at the expense of impact first investors; for example, commercial investors own equity with a higher dividend payment.
- Offering guarantees to decrease risk to commercial investors. Impact first investors can guarantee a minimum return or maximum loss to encourage participation by commercial investors.

These tactics are designed to catalyze, or crowd-in, deep-pocketed commercial investors, with the objective of proving that investments in companies such as d.light can generate attractive risk-adjusted financial returns. There are situations, however, where this strategy can backfire.

AVOIDING MARKET DISTORTIONS

Impact first investors can positively change traditional investors' perspectives on risk with respect to climate change investments in new technologies or markets. But impact first investors can also distort markets, causing more harm than good. This occurs when impact first investors finance business models that never develop a sustainable business model.

For example, in sub-Saharan Africa, nonprofit organizations have funded solar light companies to sell solar lanterns below cost as an aid to poor consumers who are otherwise unable to afford them. In some cases, the solar lanterns are even given away for free. This is laudable from a humanitarian perspective but distorts the long-term growth of the market, as companies such as d.light are unable to compete against subsidized competitors. Market distortions risk "crowding-out" commercial capital, which cannot finance enterprises lacking sustainable business models.

Impact first investors find themselves in a challenging role. They must find investment opportunities that catalyze participation by commercial investors. By definition, however, impact first investors commit capital at below-market returns on a risk-adjusted basis, which can distort a business sector and damage its long-term prospects. Therefore, to be successful, impact first investors must thread the needle between these outcomes, sourcing opportunities with both unproven business models and the potential for financial sustainability. This requires serious focus, patience, and very deep pockets.

IMPACT FIRST INVESTORS IN THE ERA OF CLIMATE CHANGE

Relatively few investors are impact first, because few are able or willing to accept below-market risk-adjusted returns. Fiduciaries, which include nearly all traditional investment funds and money managers, have a legal responsibility to maximize the risk-adjusted returns of their clients and are therefore prohibited from following an impact first investing strategy. Among individual investors, only a small number have the financial capacity to absorb significant risk and low returns over a long time period. And most individuals lack the expertise to evaluate investment opportunities, to provide support to investees in the form of training and networking, or the relationships to crowd-in commercial capital. This leaves impact first investing to a small number of people.

Given the legal and practical challenges, impact first investors are typically ultrahigh net worth, defined as those with investable assets greater than $30 million. This is an elite group, estimated at 70,000 Americans, or less than 0.03 percent of the population.[15] Ultrahigh net worth households, especially the wealthiest among them, often invest through "family offices," entities established and controlled by one or sometimes a small group of families to manage their wealth. Family offices are exempt from registration under the Securities and Exchange Commission (SEC) and have near-total discretion in how assets are managed.[16] Ultrahigh net worth families often prefer a very long-term investment horizon and can employ investment specialists in their family offices with the expertise to evaluate and support impact first investments.

Ultrahigh net worth families often establish foundations as a tax-effective means of pursuing philanthropic objectives, allowing them to make program-related investments. Even nonprofit organizations such as the Acumen Fund, which pioneered the impact first investing strategy, are funded primarily by very wealthy individuals and their foundations.[17] Fortunately, what ultrahigh net worth investors lack in number, they make up for in impact. Bill Gates (figure 14.1), one of the wealthiest Americans, is a leading example.

A HIGHER TOLERANCE FOR TECHNICAL RISK

Breakthrough Energy Ventures is the world's largest impact first investment fund focused on climate change. Founded by Bill Gates and a coalition of private investors in 2016, the fund has $1 billion to invest "in scientific breakthroughs that have the potential to deliver cheap and reliable clean energy to the world."[18] In addition to Gates, investors include Jeff Bezos, Michael Bloomberg, Ray Dalio, and more than a dozen other billionaires.

Breakthrough Energy Ventures developed a clearly defined investing strategy with four specific criteria for investment. First, the fund is focused on impact at scale, only backing technologies with the potential to reduce more than 500 million tons of greenhouse gases annually, about 1 percent of global emissions. Second, entrepreneurs seeking investment must demonstrate technology that is technically feasible at scale. Third, the fund seeks to fill in financing gaps, focusing on sectors traditional investors have avoided. And fourth, the fund seeks to back companies that will eventually attract financing from other investors, crowding-in commercial capital.[19]

Bill Gates understands the role of impact first investors, writing in his blog: "We are willing to wait a longer time for returns than other funds. We have a higher tolerance for technical risk, because we know it's tough to determine which technologies will succeed."[20] Breakthrough Energy Ventures is patient, making investments over a 20-year period, far longer than purely commercial funds. This allows the fund to back technologies that have potential for extraordinary impact yet are ignored by traditional investors.

Breakthrough Energy Ventures has backed companies with unproven technologies in solar power and energy storage, sectors in which the fund can finance cutting-edge innovations. But many of the fund's investments are in sectors that receive little attention yet hold enormous promise for addressing climate change: financing low-carbon concrete, clean fertilizers, and microgrids.[21] Most of those investments will fail to generate a financial return, but that is acceptable given the impact first investing model followed by the fund. If only one investment succeeds, it will literally move the needle on global greenhouse gas emissions, and if several succeed, the fund can claim credit for dramatically reducing the impact of climate change.

CATALYTIC CAPITAL

Breakthrough Energy Ventures is the largest pool of capital pursing an impact first strategy for addressing climate change. Fortunately, it is not alone. Other impact first investors focused on climate change include PRIME Coalition, a network of more than 150 ultrahigh net worth families, and several other smaller groups. Although an impact first investing strategy can only be used by a highly select group of extremely wealthy investors, this strategy has outsized importance.

In the context of climate change, impact first investors fill a critical role, providing early-stage "catalytic capital" to high-risk companies with unproven climate solutions. Proven technologies such as wind, solar, and electric vehicles will reduce greenhouse gas emissions by half, requiring new technologies and solutions to solve for the other half.[22] Ultimately, impact first investors are helping to mitigate catastrophic climate change by kickstarting technologies and companies that can be scaled to create a meaningful difference.

FROM STRATEGIES TO PRODUCTS

The investing strategies that have been described in section 3—managing climate risk, divestment, ESG, thematic, and impact first—provide investors with a framework for investing in the era of climate change. Next, sections 4 and 5 describe products available to investors, first in real assets and then in financial assets. Section 6 brings it all together, connecting the actions of investors to the future path of climate change.

SECTION 4

Investing in Real Assets

Real assets require large commitments of capital for development and construction, and account for most climate change investment flows. Investing in real assets requires technical expertise and deep pockets and is the domain of specialized investors. But all investors should understand the opportunities and risks of investing in real assets that offer climate solutions, as these underpin many of the financial assets described later, in section 5 of the book.

"We have got a big appetite for wind or solar. If someone walks in with a solar project tomorrow and it takes a billion dollars or three billion dollars, we're ready to do it. The more there is the better."

—WARREN BUFFETT, BERKSHIRE HATHAWAY

FIGURE 15.1. Offshore wind farm (Source: Wikimedia Commons)

Chapter Fifteen

RENEWABLE ENERGY PROJECTS

Financing renewable wind and solar projects is appealing to investors who seek reliable risk-adjusted returns. Technology risk is low, cash flows are stable and long term, and the sector is large and growing rapidly.

INVESTING IN SOLAR PROJECTS

Solar power projects generate electricity that is surprisingly predictable, allowing for accurate modeling of cash flows to forecast return on investment. Sunlight varies throughout the day with cloud cover and throughout the year with the seasons, but the total number of hours of sunlight is relatively constant year to year. The U.S. government makes available online geospatial tools and data sets on hours of sunlight, referred to as solar insolation, for anywhere in the country.[1] Understanding the different forms of solar power provides clarity on the opportunity for investors.

Sector Overview

Solar projects in the United States are of four types: residential, commercial, community, and utility. The solar panels used in all solar projects are identical, the primary difference being the size of the project.

Residential solar projects are mounted on individual homes, with the electricity generated from the solar panels used by the homeowner. The rapidly declining cost of solar panels has made residential solar projects cost-effective for many American homeowners, saving money on electrical bills even after accounting for the cost of the solar system. More than half of all American homeowners already have or are considering installation of residential solar panels.[2] For investors, however, residential solar projects are too small to finance individually. Instead, solar leases and loans are bundled together to create investment vehicles, explained in section 5.

Commercial solar projects supply electricity to buildings or businesses. Like residential solar projects, most commercial-scale projects are located on building rooftops, although projects are occasionally located on land next to the building. Again, like residential solar, commercial solar projects generate electricity that is used on-site. Building owners install commercial solar projects for the same reason as homeowners, to save money on electrical bills. And like residential solar, projects are financed through investment vehicles, explained in section 5.

Community solar is a relatively new form of solar project, generating electricity that is shared by multiple homeowners and businesses. Instead of placing solar panels on homes or business properties, a community solar project is constructed on leased land, and the electricity generated is sold to the utility for use on the power grid. The economic benefits of the project are shared by the members who own it. Community solar is attractive to the many U.S. residents and businesses that cannot otherwise host a solar project, those living in multi-tenant buildings or with roofs that are unable to support a solar system. Community solar is increasingly popular, but it does not offer investors outside of the community the opportunity to participate.

Investors seeking real asset investment opportunities focus on utility-scale projects. These are large in scale and generate electricity for distribution to the electrical grid. To provide perspective, the smallest utility-scale solar projects are composed of approximately 4,000 panels on 5 acres of land, and the largest utility-scale projects stretch over thousands of acres of land. For example, the Samson Solar project under construction in Texas will install several million solar panels and generate enough power for a city of 300,000 homes.[3] Utility-scale projects account for more than three-quarters of new solar generation constructed in the United States, creating demand for hundreds of billions of dollars in project investment capital.[4] Institutional investors such as Warren Buffet's Berkshire Hathaway are the primary source of financing given the large size and capital needs of each project. Smaller investors participate in utility-scale solar projects by placing capital in funds that own project portfolios, described in chapter 21.

Investing in Utility-Scale Solar Projects

Investors in utility-scale solar projects model five key project attributes—capital costs, operating costs, electricity generation, the power purchase agreement, and government incentives—to calculate investment returns, as follows:

- *Capital costs* are incurred for solar panels and the balance of systems, which includes mounting racks to hold the panels, an inverter to convert electricity generated from direct current (DC) to alternating current (AC), electrical cables, and monitoring equipment. Some projects also install a tracking system on the mounting racks, which allows the solar panels to follow the arc of the sun. Utility-scale solar projects must be connected to the electrical grid, which can require construction of transmission lines and installation of equipment for the interconnect. Engineering, procurement, and construction (EPC) costs include labor for installing panels and the balance of systems. Solar projects also incur soft costs, which include the cost of sourcing and contracting with the site owner, permitting, overhead, and margin.
- *Operating costs* are low. Solar PV systems require little maintenance as there are no moving parts, other than a tracking system if installed. The primary operating cost is siting, the use of the land on which the solar panels are located. Lease payments for the use of land can be significant, ranging from $250 to $2,000 per acre, as the land cannot be used simultaneously for farming or other purposes.[5] Other operating costs typically include insurance, administration, and property taxes.
- *Electricity generation* is forecast over the life of the project, typically assumed to be 25–30 years, although solar PV panels can operate even longer. Solar panel power capacity is guaranteed by panel manufacturers and degrades at approximately 0.5 percent per year.[6] Electricity generation is a function of total solar panel power capacity multiplied by the amount of sunlight or solar radiation reaching the PV panels (insolation).
- *Power purchase agreement* (PPA) is a long-term fixed-price contract between the owner of the solar project and a utility or large business to purchase the electricity generated. PPAs allow investors to lock-in revenues over the life of the project, as solar project revenue is simply electricity generation multiplied by the PPA price. For investors, solar projects normally have a long-term PPA with a creditworthy counterparty. PPAs typically are 10–25 years to match the long-term generation capacity of solar projects.[7]
- In the United States, *government incentives* for solar projects provide an important component of project economics and primarily take two forms: tax incentives and renewable portfolio standards, described below.

Incentives in the Federal Tax Code

The U.S. federal tax code offers tax credits to encourage investment in infrastructure projects, especially in the energy industry. Solar projects benefit specifically from the investment tax credit (ITC), authorized by Congress to provide investors with an incentive to finance solar projects. The ITC provides a dollar-for-dollar reduction in income taxes that an investor would otherwise pay the U.S. government. As of 2022, the ITC is 26 percent of the capital cost of projects commencing construction by 2023, declining to 22 percent in 2024 and 10 percent thereafter.[8]

Solar projects are also eligible for the federal tax benefits of accelerated depreciation. With modified accelerated cost recovery system (MACRS), the capital cost of a solar project can be depreciated over 5 years, irrespective of the actual operating life of the project. Investors in solar projects use MACRS to reduce taxable income and improve after-tax return on investment.

Solar projects located in the United States can attract capital on favorable terms from tax equity investors, a niche financing sector that provides capital in return for the tax credits and accelerated depreciation from the project. Traditional equity investors in solar projects benefit by co-investing with tax equity investors, earning higher returns as the tax equity is nondilutive.

Renewable Portfolio Standards

Thirty U.S. states have enacted renewable portfolio standards (RPSs), regulations that require utilities operating in a state to source a minimum percentage of electricity from renewable power sources, with a penalty for noncompliance.[9] For example, New Jersey's renewable portfolio standard mandates utilities procure 22.5 percent of electricity sold in the state from renewable energy, with a specific requirement that a minimum of 4.1 percent be from solar power.[10] RPS regulations incentivize development of solar projects by creating additional demand for renewable energy.

States with a renewable portfolio standard issue renewable energy certificates (RECs) to solar projects, each REC certifying generation of 1 megawatt-hour of electricity from renewable energy.[11] RECs are a useful way to track energy sources because electricity generated from renewables is indistinguishable from that generated by fossil fuels. In some states, certificates are called SRECs (*solar* renewable energy certificates) to distinguish them from RECs generated from other renewable sources.

RECs improve the financial return on solar projects by creating an additional revenue stream when sold to utilities to meet RPS compliance. But investors in solar projects need to be aware that REC/SREC prices vary widely by state and by year. For example, in 2021, SREC prices in New Jersey traded for $230 and in Massachusetts for $322,[12] yet a decade earlier, SRECs generated in New Jersey sold for more than $600.[13]

Investment Risks

Development of solar projects is a high-risk, high-return business and requires specialist knowledge for sourcing sites and land leases, negotiating PPAs, securing interconnect agreements, and navigating regulatory and community approval processes. Project development is risky because a failure at any step in the process will bring the project to a premature end with little or no residual value. But once a project is constructed and operating, referred to as the commercial operation date (COD), the risk of failure is very low. Solar project developers typically sell projects that reach COD to long-term institutional investors, using the proceeds to develop new projects.

For investors in operating utility-scale solar projects, the risks are low. PV panel risks are minimal, as panel performance is guaranteed by the manufacturer, generally for up to 25 years.[14] Inverter performance presents a slightly greater risk, as inverters are often replaced after a decade.[15] Contractual risk exists if the PPA counterparty, the buyer of electricity from the project, declares bankruptcy or attempts to renegotiate the contract; however, this is very rare because PPAs are predominantly with large, creditworthy utilities or corporations. Financial risk exists with REC prices, as most projects cannot secure contracts beyond 5 years to sell RECs, but REC cash flows generally account for a modest percentage of total project cash flows. In fact, the biggest risk for investors in solar projects is exogenous to the project: exposure to rising interest rates.

Solar projects act like fixed-income securities, as cash flows are stable and long term. This is advantageous when interest rates are falling, as the value of a solar project will increase when interest rates decline. Conversely, project asset values fall when interest rates are rising. In this way, the risk to solar projects is very similar to that of most bonds and other fixed-income securities. For investors in solar projects, exposure to interest rate risk should be evaluated as it would be for any investment in assets with fixed cash flows.

"Rock-Solid" Solar

Investors seeking long-term, low-risk returns are increasingly attracted to the solar sector. Analysis of the sector by Fitch Ratings found "really rock-solid performance" with 93 percent of projects performing at or above projections.[16] Given the steady performance, it should come as no surprise that investment returns are modest. Project IRRs range from 6 to 8 percent, with some assets yielding less than 6 percent because of surging demand from pension funds and insurance companies that view solar projects as a stable asset.[17] Strong investor demand is also occurring in a related sector: wind power.

INVESTING IN WIND PROJECTS

Wind power offers a similar investment opportunity as solar, but the technology and government incentives are slightly more complex. Like solar, investing in wind projects requires careful modeling of cash flows to determine investment returns. Unlike solar, wind projects are almost always utility-scale, using enormous wind turbines to generate power for the electrical grid (figure 15.1).

Evaluating Project Economics

The economics of building and operating a wind farm are determined primarily by the capital costs of the wind turbines and the installation and interconnection to the grid to transmit the electricity generated by the wind farm. The largest capital cost is the wind turbine itself, which is composed of a tower, a nacelle (which houses the equipment at the top of the tower), and the blades. Installation of a wind turbine is straightforward, although transport of the turbine from the factory to the wind farm site can be challenging because of the immense size of the various components. An advanced wind turbine is significantly larger than a jumbo jet, so special-purpose ships and trucks are required to transport the turbines.

Once at location, the wind turbine is placed on a steel-reinforced cement foundation and connected to the electrical grid, the latter called an interconnection. The cost of the interconnection is mostly dependent on the distance between the wind farm and the nearest transmission lines. This can be costly for wind projects, as the windiest places in the United States—the Dakotas, Nebraska, Kansas, and West Texas—are far from population centers where demand for electricity is greatest.[18]

A wind turbine has a rated power output in megawatts (MW), which represents the maximum amount of power that can be generated by the turbine at any given point in time. The electricity generated by a wind turbine is the rated power output in megawatts multiplied by the number of hours of operation. Of course, wind turbines only generate electricity when the wind is strong enough to turn the blades, typically a wind speed greater than 7 mph.[19]

The most important calculation for forecasting the cash flows of a wind farm is the capacity factor, defined as the ratio of actual power output to rated power capacity if the turbines were always operating. Onshore wind farms, located on land, experience average capacity factors of 35 percent.[20] In other words, the wind is strong enough to turn the blades an average of 35 percent of the turbine's rated output throughout the year. For example, a 5-MW wind turbine operating with a 35 percent capacity factor will generate 15,330 megawatt-hours (MWh) of electricity per year.[21] Note that offshore wind projects experience higher capacity factors, as wind speeds are typically higher and steadier over the water.

Unfortunately, forecasting wind power is more challenging and less accurate than forecasting solar insolation, as wind power output varies as a function of a cube of the wind speed, whereas solar power output has a linear relationship to solar insolation.[22] The U.S. government offers public wind maps, but wind project developers require extremely accurate measurements of local wind speeds to forecast power generation and to determine the precise siting of turbines to minimize turbulence in wind flow from one turbine to the next. (Figure 15.2 shows how turbines affect wind flow.) In most cases, project developers will install an anemometer on the proposed site to measure wind speeds for at least a year before committing investment capital.

The operating costs of a wind farm are primarily the costs of leasing the land on which the turbines are located, operating the project, and maintenance. Operating a wind farm requires little oversight or expense, as wind turbines are highly automated. Maintenance, however, can be costly. Wind turbines are designed to generate electricity for more than 20 years, but they are mechanical devices, which means that components eventually wear out and must be replaced.

FIGURE 15.2. Wind farm operating on a foggy day, demonstrating how wind flows can be affected by turbines (Source: Vattenfall Horns Rev project, Denmark)

Farming the Wind

Wind farms are often sited on agricultural land, generating incremental income for farmers who lease acreage to the wind project owners. Each turbine requires approximately 50–100 acres of land, as the turbines must be spaced far apart to avoid wind turbulence from one turbine affecting the performance of others.[23] Despite their large size, the base of a wind turbine is small, so even the largest wind turbines require less than an acre of land for the foundation. This means farmers can continue to use land they have leased to wind farms for grazing livestock or growing crops. In the United States, farmers are typically paid $7,000 to $10,000 annually per turbine for leasing the land to the wind turbine owner,[24] dramatically more revenue than that from farming the equivalent land area.[25] Earnings from wind turbines provide a buffer against fluctuating agricultural commodity prices, as leases are fixed for 20 years or more. Additionally, the increase in land values from wind projects results in more tax revenue for state and local municipalities. According to a farmer in Iowa "this is our financial future."[26]

PPAs and Virtual PPAs

Wind projects transmit power directly to the electrical grid, allowing project owners to sell electricity at the spot wholesale rate or at fixed rates by entering into a power purchase agreement (PPA) contract. As with utility-scale solar projects, PPAs provide wind investors with long-term revenue certainty. Wind project PPAs typically fix rates for 15–25 years, securing revenue for the life of the project and significantly reducing project risk.[27] But securing a PPA with a utility can be challenging because many utilities are highly regulated and have little need for additional power supply. An alternative for wind (and solar) projects is the *virtual* PPA (VPPA), a financial contract with a large corporate buyer to exchange the variable wholesale price for a long-term fixed price. VPPAs allow companies to hedge the cost of electricity and provide wind and solar project owners with fixed rates, benefiting both parties.

For investors, wind and solar projects secured with long-term VPPAs can dramatically reduce risk, as counterparties are typically of high credit quality. Amazon has entered into multiple VPPAs to procure wind and solar power across the country, and it is not alone. Starbucks, Microsoft, McDonalds, and The Home Depot, to name just a few, all signed VPPAs in 2020.[28] A study by Columbia University's Global Energy Policy Center found that VPPAs are an increasingly popular structure to simultaneously hedge power prices and support renewable energy development, and they expect significant growth in the years ahead.[29]

Government Incentives

The federal government has incentivized wind project development with production tax credits (PTCs), allocated on a per kilowatt-hour (kWh) basis for the first 10 years of operation. In 2020, the PTC was reduced to $0.018 per kilowatt-hour, and the entire program expired at the end of 2021, although new legislation may extend it.[30]

Offshore wind projects in U.S. waters that begin construction before 2026 are eligible for a 30 percent investment tax credit, like the incentives received on solar projects.[31] The offshore wind industry is nascent in America, with a relatively high levelized cost of electricity (LCOE), making it uncompetitive with other forms of electricity generation. The investment tax credit for offshore wind projects is designed to boost the number of projects under development, which should reduce costs and make offshore wind a viable long-term source of power.

Wind projects are eligible, like solar, for accelerated depreciation (MACRS), allowing depreciation of most capital costs over 5 years irrespective of the project's actual operating life.

State governments include wind power in renewable portfolio standards (RPSs). However, the price of wind renewable energy certificates (RECs) is low in many states because of rapid development of wind projects across the country. In Texas, the top wind energy–generating state in the nation, wind supplies 23 percent of the grid's power.[32] Unsurprisingly, the price of a wind REC in Texas is very low, trading around $1 per megawatt-hour,[33] a fraction of the $22 to $38 per megawatt-hour wholesale price for electricity.[34] Low wind REC prices are proof, as if more is needed, that wind in America is competitive with all other sources of power generation.

Green Banks

In addition to tax and regulatory incentives, 14 states have sponsored *green banks* to facilitate investment in renewable energy projects that have financing gaps not currently well-served by the private sector. Green banks are impact first investors (described in chapter 14) with a focus on infrastructure. The objective of green banks is to invest early in nascent sectors to demonstrate to investors that risk is low, encouraging commercial financing of subsequent projects.

New York State established the country's largest green bank in 2014, committing $1.1 billion with the objective of catalyzing $3 billion in private sector clean-energy investments.[35] In the wind sector, New York Green Bank is supporting the state's plan for development of 9 gigawatts (GW) of offshore wind, where commercial investors have less experience. By using innovative wind-financing structures to reduce risk, New York Green Bank can entice commercial investment capital into a new and growing segment of the wind sector.

Investment Risks

Investing in wind projects requires an understanding of potential risks: mechanical failure, wind droughts, and extreme weather. The first modern wind turbines experienced frequent mechanical failure, especially in the gear boxes, contributing to high repair costs. Subsequent improvements in engineering design led to gearless turbines, which are more reliable and therefore less costly to maintain than earlier models. Modern wind turbines achieve an availability rate of 98 percent but still fail at least once per year.[36]

Extreme weather, including violent storms and lightning, can damage wind turbines. Wind farms generate more power as the wind speed increases, up to a point. Beyond the cut-out speed, typically 55 mph, turbines automatically shut off by feathering the blades.[37] But the biggest risk to wind investors is not too much wind, but too little.

Accurately predicting wind speed, and therefore electricity generation and revenue over the life of the project, is essential to forecasting investment returns. Unfortunately, it is difficult to do this well. Wind project engineers calculate a figure called P50, the annual electrical generation projected to be exceeded 50 percent of the time. P50 is determined using on-site wind measurements collected over a minimum of 1 year prior to project development. In theory, P50 provides an accurate forecast of wind and electricity generation. In fact, Fitch Ratings has found that "only 24 percent of wind project observations were within 5 percent or better of the original P50 levels," due to wind variability or poor siting of turbines. Given that risk, investors typically conduct sensitivity analysis of project cash flows for lower than forecasted wind speeds.

"Investors Have Learned to Love Wind"

The *Wall Street Journal* ran that headline in 2020, in an article that summed up the benefits of wind projects for investors: steady, long-term income and returns in the high single digits.[38] Investment IRRs of 6–8 percent are common for low-risk operating projects sited in North America, with equity investors increasing returns by adding financial leverage.[39] Investing directly in wind-project assets is slightly riskier than in solar projects given the greater variability in wind speed but has the advantage of scale, as wind projects are typically much larger than solar projects, providing an opportunity for large institutional investors to place capital in a relatively safe asset that generates attractive risk-adjusted returns.

The intermittency of wind and solar projects—the fact that these renewable resources generate power only when it is windy or sunny—is giving rise to a related real asset investment opportunity: energy storage.

ENERGY STORAGE PROJECTS

The rapid growth in renewable solar and wind projects is, unsurprisingly, driving demand for energy storage projects. In the United States, pumped hydro projects have historically been the preferred technology for energy storage, but the vast majority of new projects use batteries. Battery energy storage systems (BESSs) have many benefits: flexible configurations, rapid charge and discharge, high power output, and no maintenance. And battery systems are increasingly attractive as costs decline.

The sector is nascent. Investment in the United States totaled a mere $6 billion in 2021, but it is almost certain to expand alongside the growing challenge of intermittency from renewable power on the electrical grid.[40] For investors, energy storage projects offer higher returns than either wind or solar projects because investing in storage is significantly more complex and riskier.

Creating Value with Storage

Unlike solar and wind power, which create value by generating electricity, energy storage can create value in multiple ways, each of which can potentially be monetized by project owners. The most common applications include the following:

- Short-duration storage: Energy storage of less than 6 hours creates value by providing electricity to the grid when intermittency from low-wind and cloudy days affects renewable power generation.
- Peaking capacity: The U.S. electrical grid relies on more than 1,000 peaker plants to generate electricity on days with high demand for power, especially during hot summers when air conditioners are widely used.[41] Most peaker plants in America are powered by natural gas and are costly because they are often used just a few days per year.[42] Cost-competitive energy storage projects can replace peakers, saving utilities from having to invest in and maintain little-used facilities.
- Long-duration storage: Energy storage of more than 6 hours of power creates value for long duration events, providing electricity to the grid during multi-day storms and months with low wind or continual cloud cover.

- Energy shifting: Energy storage systems can purchase and store power when electricity generation is in surplus in the middle of the night and sell power when demand is high during the day. Energy shifting allows for optimal use of wind and solar energy while arbitraging the difference in wholesale power prices between periods of high and low demand.
- Transmission and distribution upgrade deferrals: Costly upgrades to the grid can be delayed or avoided with energy storage systems. Communities with electrical grids operating at maximum capacity often face expensive upgrades when demand increases. Energy storage projects can delay or eliminate that need by providing additional power at nodes on the grid close to customer demand.

Government Incentives

Battery energy storage systems are eligible for the federal investment tax credit (ITC) but only if the system is paired with a renewable energy project. Legislation has been proposed to extend the ITC to stand-alone projects. Like solar, the ITC is 26 percent of the capital cost of projects commencing construction by 2023, declining to 22 percent in 2024 and 10 percent thereafter. Energy storage is also eligible for accelerated depreciation (MACRS) on a 7- or 5-year basis depending on how much of the project is paired with renewables.

States offer a range of incentives for energy storage projects, from property tax incentives to storage targets. Similar in design to renewable portfolio standards (RPSs), storage targets mandate that a minimum capacity be developed by utilities operating in the state. In 2013, California established the first U.S. energy storage target, requiring the state's three investor-owned utilities to develop 1,325 MW by 2020, equivalent to approximately five peaker plants. California's utilities met the target 1 year early, and energy storage projects in the state lead the nation.[43]

Better Together

Developers of energy storage projects find that financial returns can be optimized when storage is combined with solar and wind projects. Renewable power generation creates value for energy storage through load shifting and peaking capacity, while energy storage creates value for wind and solar by providing short-duration storage. Using this strategy, NextEra Energy Resources, the largest owner of renewable energy assets in America, is developing a "triple hybrid" project in Oklahoma that includes 250 MW of wind, 250 MW of solar, and 200 MW of 4-hour battery storage.[44] NextEra announced: "The combination of low-cost renewables plus storage is expected to be increasingly disruptive to the nation's generation fleet, providing significant growth opportunities well into the next decade."[45]

For investors, financing energy storage, especially when combined with wind and solar, looks like a sure thing. But there are significant risks specific to energy storage that need to be considered when analyzing project investment opportunities.

Energy Storage Risks and Returns

As described above, energy storage projects can add value in many ways, generating multiple sources of revenue. However, there is no standard revenue model for energy storage projects, which lack the long-term PPA contracts of renewable wind and solar projects, exposing investors to the risk of lower prices and revenue in the later years of the project.

Investors in energy storage projects also face technology risk. Unlike wind or solar projects, battery energy storage systems are still at an early stage of development, with several different battery technologies competing to provide solutions, including lithium-ion, redox and vanadium flow, and zinc-air, among others. In general, lithium-ion provides a better short-duration solution, while flow batteries are superior at long-duration storage. Selecting the optimal technology for an energy storage project remains challenging.

Investors are also exposed to battery performance risk. Warranties provided by battery manufacturers are typically only 2 years, much shorter than solar and wind company products, although extended warranties out to 10 years can be purchased.[46] Batteries also degrade relatively quickly as the number of times a battery is cycled affects its longevity. Lazard estimates 2–3 percent per year versus 0.5 percent for solar panels, so project investors must assume replacement of batteries when modeling financial returns.[47]

Siting can be surprisingly difficult given that even large BESSs require no more than a few acres of land. New York City banned the use of battery systems prior to 2020 because of fire risk, and many communities can be slow to approve storage projects, reducing investment opportunities and returns.[48]

Obtaining loans on battery energy storage systems can be difficult, or only at high interest rates, as projects may be considered "unbankable" because of technology risk and limited warranties.[49] Low financial leverage on a project reduces equity returns. But with higher risks come greater rewards, and the energy storage market is no exception. Investment bank Lazard reviewed energy storage projects across several applications and estimated equity returns ranging from a low of 8.1 percent to a high of 33.7 percent.[50] Returns are likely to come down as the sector matures.

Looking Ahead

The rapidly declining cost of energy storage, combined with extraordinary growth in intermittent solar and wind power projects, ensures a large market in the years ahead. Ryan Brown, CEO of energy storage company Salient, summed up the opportunity: "The best is definitely yet to come. We know that the industry is still in its infancy in almost all respects. While adoption is already meaningful and rapidly accelerating, a clean energy world will require trillions of dollars' worth of additional capacity to be installed. The 2020s will be a breakout decade for the energy storage sector."[51]

Bloomberg estimates the global energy storage market will grow 122-fold in the next two decades, requiring $662 billion in investment capital.[52] For investors, energy storage will likely become an increasingly attractive real asset financing opportunity alongside solar and wind projects.

INVESTING IN RENEWABLE ENERGY AND STORAGE PROJECTS

Investing in renewable energy projects can provide stable financial returns, as evidenced by the performance of wind and solar projects through the COVID-19 pandemic. According to Fitch Ratings, "The ongoing global pandemic has had a muted impact on this class of credits that are largely insulated from demand risk and have remained operationally stable."[53]

Wind and solar projects in the United States are a $50 billion annual market, and the sector is poised to grow with the addition of energy storage.[54] In the future, hybrid projects that combine wind, solar, and storage will be increasingly competitive with traditional coal and natural gas plants, providing dispatchable power at cost-effective rates. For real assets investors, wind, solar, and energy storage will continue to provide attractive risk-adjusted opportunities in the decades ahead.

"If we can prove it works here, then it can work anywhere"

—DANA ROBBINS SCHNEIDER, SVP,
EMPIRE STATE BUILDING

FIGURE 16.1. Empire State Building, New York City (Source: Wikimedia Commons)

Chapter Sixteen

REAL ESTATE

Constructed nearly a century ago, the iconic Empire State Building is now, improbably, at the forefront of investment in the era of climate change. In 2010, the building owner completed an energy retrofit at a cost of $13 million to refurbish windows, install LEDs, and add regenerative braking in elevators. Over a decade, the investment has saved $4.4 million annually on energy costs, generating a payback in only 3 years.[1] The retrofit also slashed the building's greenhouse gas emissions by 40 percent.[2]

Tony Malkin, CEO of the Empire State Realty Trust, which owns its namesake building, found that the energy retrofit both reduced operating costs and netted higher rents from tenants.[3] Malkin was so impressed with the results that he announced a second project aimed at reducing emissions a further 40 percent.[4] The Empire State Building may be unique among American properties, but the financial savings on energy efficiency enjoyed by its owners are a widespread investment opportunity.

ATTRACTIVE RETURNS, DIFFICULT TO CAPTURE

Buildings are responsible for 13 percent of all greenhouse gas emissions in America; unfortunately, buildings are long-lived assets that require sizeable investments to upgrade or replace.[5] An analysis of 50,000 buildings in New York City found that compliance with local laws aimed at reducing emissions will require $20 billion in capital for commercial buildings alone.[6]

Fortunately, improvement of energy efficiency in buildings through the use of proven technologies can generate attractive investment returns at low risk. John Mandyck of the Urban Green Council described the opportunity to reduce building emissions in New York and other large cities: "The pathway to a low-carbon future is literally standing in front of our eyes."[7]

The attractive investment returns from energy efficiency projects result mainly from the fact that it is less costly to save power, referred to as a "negawatt," than it is to generate a watt of additional power. In large buildings, retrofits of roofs, windows, and doors can reduce heating and cooling demand by 40 percent, and simply replacing incandescent lights with LEDs reduces energy for lighting by 80 percent.[8] Sophisticated building automation and control systems can further reduce costs. Regrettably, building owners often find that capturing those cost savings is a greater challenge than the upgrade itself.

The triple net lease, a common arrangement in commercial buildings, passes all utility costs to the tenant. This creates a problem, as energy efficiency projects are financed by landlords while the savings are captured by the tenants. Owners are also reluctant to upgrade buildings that may be sold before the efficiency improvements pay back, as buyers can be reluctant to factor lower energy costs in the purchase price. Given these challenges, landlords are turning to a financial product called C-PACE to overcome the barriers to efficiency upgrades.

FINANCIAL INNOVATION

Commercial property assessed clean energy (C-PACE) is a recent adaptation of a financing tool used for decades by state and local governments. To encourage energy efficiency, governments allow building owners to finance upgrades with up to 20-year loans repaid through their local property tax bill. Interest rates on C-PACE loans are very low because the tax assessment mechanism is secure and very low-risk, improving the returns on energy efficiency projects. Importantly, the tax assessment transfers with the property upon sale, eliminating the risk of selling a building with an outstanding loan. Thirty U.S. states have authorized C-PACE financing, which is used by both small building owners and large companies such as shopping mall owner Simon Property Group.[9]

Investing in building upgrades and energy efficiency is an attractive prospect for real estate owners to improve financial returns and reduce CO_2

emissions. But a much greater opportunity lies in the management of risk associated with climate change and physical assets.

CLIMATE RISK

Rising temperatures are raising global sea levels because of thermal expansion and are increasing the frequency and intensity of wildfires because of drier forests, existential threats to homes and buildings located in at-risk areas. Understanding climate risks to real estate can improve investor returns and reduce potential losses.

Real estate investors face both physical risks and transition risks. Academic studies find that homes exposed to rising seas sell for 7 percent less than equivalent properties,[10] with projected price impacts of 15–35 percent by 2050.[11] Real estate data firm Zillow estimates 800,000 U.S. homes worth $451 billion are at risk of flood inundation by 2050.[12]

Wildfire risk increases with climate change as rising temperatures dry out forests and reduce precipitation. Columbia University scientist Park Williams explains the challenge posed by wildfires: "The ability of dry fuels to promote large fires is nonlinear, which has allowed warming to become increasingly impactful."[13] The National Climate Assessment predicts that annual acres burned in the western United States could increase by 2–6 times current losses by 2050.[14]

Investors in real estate and other real assets often mistakenly conclude that because the physical risks to their properties are decades in the future, the risk to their assets is similarly distant. But *transition risks* can create an immediate threat to asset prices. Real estate investors can find themselves exposed to losses decades before their properties face the physical risks of climate change.

A VICIOUS CYCLE

Transition risks occur when property owners exposed to climate change are unable to finance or insure their properties or can do so only at exorbitant rates.[15] This can set off a vicious cycle of declining property values, lower tax revenues, fewer buyers, and less financing, driving prices even lower.

Mortgages are a weak point as the traditional 30-year mortgage for homeowners puts lenders in flood-prone areas at risk of declining collateral values before those loans are repaid. Research of bank activity finds them shifting mortgages with flood risk from their balance sheets to Fannie Mae and Freddie Mac. Jesse Keenan, an associate professor at Tulane University, warns that "conventional mortgages have survived many financial crises, but they may not survive the climate crisis."[16]

Insurance is the other weak point, as insurers stop offering coverage in fire-prone counties to avoid exposure to the growing threat of wildfires. California, which has experienced several of the country's worst forest fires, is experiencing this climate transition risk. The state's insurance commissioner summarized the challenge: "By not being able to find insurance, you then in turn can't sell your home. If you can't sell your home, then it affects the local property taxes. This is really creating chaos."[17]

Transition risks are the market's expectation that physical risks will manifest themselves within the life span of the asset. In real estate, transition risks include exposure to sharply higher mortgage and insurance rates, property damage from nuisance flooding and extreme weather events, and rising property taxes to pay for local community repairs and resilience. Added to this burden is a decline in liquidity, as home buyers and employers decide to avoid at-risk locations.

Real estate investors in communities at risk are hoping government will come to the rescue.

INFRASTRUCTURE HARDENING

Municipalities exposed to flooding or wildfires are making plans to address these risks to their communities. Miami Beach is raising roads in areas that flood frequently, and New York City is designing a storm surge barrier wall to protect against a rising sea.[18] The New York project is estimated to cost $119 billion and to take an estimated 25 years to build.[19] In wealthy communities with concentrated real estate, this might make economic sense. But for much of the country, government spending on adaptation to climate change is both costly and likely to be ineffective, a stopgap at best.

Infrastructure hardening is challenging because climate change may be moving faster than government-funded construction. A case in point is Venice, Italy, a low-lying city of great economic and cultural value that in 2003 had the foresight to begin construction on a flood barrier designed for 8 inches of sea-level rise. Unfortunately, since the project began, projections for sea-level rise have increased to 14 inches by 2100. Professor Rusconi of Venice University describes the flood barrier as a "costly, useless project with no guarantee it will function well."[20] Critics of the proposed New York City barrier project also highlight that the government's plan may be obsolete within decades because of the relentless pace of climate change.

Governments the world over will find it challenging to invest in long-term solutions to climate change given the extremely high cost of infrastructure hardening and the uncertain payoff. For many homeowners and commercial tenants, it is simply easier to move.

LONG-LIVED ASSETS MEET A LONG-TERM THREAT

The risk that climate change poses to real estate is neither new nor controversial. The chief economist at mortgage giant Freddie Mac warned in 2016 that rising seas "appear likely to destroy billions of dollars in property."[21] For real estate investors, the challenge is not whether climate change will erode asset values, but when. Building stocks are long-lived assets, turning over only 1–3 percent per year.[22] Investing with a long-term perspective has traditionally been considered a competitive advantage. No longer. Climate change is now a long-term threat. Real estate investors inclined to commit for the long term may find themselves holding assets that rapidly decline in value.

There are no easy solutions for protecting real estate in the era of climate change. Savvy real estate investors are recognizing the near-term transition risks and longer-term physical risks from climate change and are pricing those into asset valuations. For many, the best strategy may simply be to sell at-risk assets in exchange for climate-resilient properties.

"Even if some farmers don't believe in climate change, they believe in money."

—SUCCESSFUL FARMING MAGAZINE

FIGURE 17.1. Hay bales being collected (Source: Wikimedia Commons)

Chapter Seventeen

FORESTRY AND AGRICULTURE

The world is losing 40 million acres of forest cover each year, primarily driven by deforestation to support agriculture.[1] From cattle ranching in Brazil to palm oil plantations in Indonesia, the losses are greatest in tropical forest countries with rapidly expanding populations and economies.[2] This creates an impact on climate change by reversing the biological process by which trees and other plants capture and store CO_2, as clearing land releases carbon previously sequestered in trees and destroys the CO_2-absorbing capacity of the forest. Globally, deforestation is responsible for 11 percent of greenhouse gas emissions.[3]

Investing in forestry projects to reverse the damage wrought by deforestation offers a scalable, inexpensive climate change solution. A study by researchers at Yale and Harvard found that the cost of reducing CO_2 through forestry is $1 to $10 per ton, dramatically less than most other climate solutions.[4] And investments in forestry could reduce CO_2 by nearly 8 billion tons per year in 2030, more than 20 percent of global emissions.[5]

Forestry projects take two forms: protecting trees from being cut down (*avoided deforestation*) and planting trees (*reforestation* and *afforestation*). Both project types are scalable and inexpensive. But there is a significant hurdle faced by investors in forestry projects: securing payment for reducing emissions. Carbon markets can solve that problem.

CARBON MARKETS IN THEORY

The economic theory upon which carbon markets are based was first postulated by an obscure Canadian economist, John Dales, in 1968. Dales recognized that it is extremely difficult for any government to accurately establish a price on emissions, as costs to reduce emissions constantly change with new technologies, innovation, and market conditions. Without knowing the cost to reduce emissions, it becomes nearly impossible for a government to assign a price on emissions, and without a price, businesses are unwilling to invest to reduce pollution.

Fortunately, Dales had an elegant solution, which was to place a cap on the overall level of pollution and assign rights allowing businesses to collectively pollute up to the cap but no more. Importantly, businesses would be allowed to trade their rights to pollute, creating a market. Dales wrote "the virtues of the market mechanism are that no person, or agency, has to set the price—it is set by the competition among buyers and sellers of rights."[6]

In a carbon market, capped participants decide whether to reduce emissions directly themselves or to purchase rights to emit from other participants. Capped participants will trade among themselves to optimize their position if the cost of reducing emissions is below the penalty price imposed by the government for noncompliance. In this way, the invisible hand of the market finds the least costly emissions reduction technology from among any number of options and incentivizes businesses to constantly find new and less costly solutions, creating savings for the entire market.

FROM THEORY TO PRACTICE

Dales's theory was not put to the test until 1990, when President George H. W. Bush supported an amendment to the Clean Air Act that allowed for the establishment of a cap-and-trade market to address emissions from coal-fired utilities that caused acid rain. Utilities facing a cap on emissions could reduce pollution directly or by purchasing a right to emit from another utility that reduced emissions at lower cost. This market created a financial incentive for polluters to reduce emissions as quickly and deeply as possible because they could sell excess rights, thereby generating revenue in return for lower emissions.

The acid rain cap-and-trade program was a stunning success, reducing emissions by 50 percent at very low cost to business.[7] More important, the program proved that a market-based system was capable of capping emissions at a desired level, reducing pollution at the lowest overall cost to the country. The dramatic success of the U.S. acid rain program encouraged economists working on climate change to design a similar system for reducing greenhouse gas emissions, which came to be called carbon markets.

CARBON MARKETS AND PROJECT INVESTMENT

Carbon markets are designed by governments to reduce greenhouse gas emissions *at lowest cost*, using the market to set the price of emissions and thereby guide investment decisions. This creates an efficient allocation of capital to climate solutions such as forestry, which have the capacity to reduce emissions inexpensively when compared to other emission reduction solutions.

Deforestation occurs because timber can be sold, and cleared land can be used for agriculture, generating revenue for the landowner. Carbon markets create an alternative source of revenue that can make trees more valuable standing than cut down, incentivizing investment to protect forests. Projects that reduce greenhouse gases from entering the atmosphere or remove them from the atmosphere have the potential to earn carbon credits. A carbon credit represents 1 metric ton of CO_2 equivalent. Note that carbon market terminology can be confusing—in this book, rights to emit, allowances, credits, and offsets are all referred to as carbon credits.

Selling carbon credits provides a clever way to earn revenue in return for protecting forests. But investors in forestry projects designed to generate carbon credits need to be aware of several specific challenges:

- *Additionality*: To be eligible for carbon credits, projects must demonstrate additionality, meaning the project is either protecting trees that will be deforested in the absence of the project, or planting trees faster than would occur naturally. Additionality is necessary in carbon credit projects to ensure that the credits generated are contributing to a reduction in greenhouse gas emissions, as those credits will allow a polluting company to emit CO_2 elsewhere. Investors in forestry projects must be confident that the project will pass the additionality test, as failure to prove additionality prohibits the project from earning any carbon credits, resulting in a total loss of revenue for the project.
- *Impermanence*: Even protected forests can be destroyed by wildfires, disease, or insects, releasing stored carbon into the atmosphere and reversing the positive

climate impacts of a forestry project. And this risk may increase as global warming puts stress on the natural environment.[8] Investors must account for the probability of events that reduce forest permanence and factor that into the project's projections for CO_2 sequestration, thereby reducing the projected number of carbon credits to be issued.

- *Leakage*: Projects that protect forests in one place can result in an unintended increase in deforestation in another location, eliminating the climate change benefits of the project. Investors must evaluate the risk of leakage from the project and account for this in the determination of carbon credits generated by the project.
- *Verification*: The carbon captured and sequestered in a forestry project must be accurately verified by a credible auditing firm for the project to receive credits. This is normally done annually. Verifying forest emissions can be challenging because physically measuring and counting trees is unrealistic. Instead, verifiers use sophisticated satellite imaging combined with airborne LiDAR measurements from planes or drones to determine forest carbon stocks.[9]
- *Prices*: Carbon credit prices fluctuate by market and over time with changes in supply of and demand for credits. This creates risk for investors in forestry projects lacking long-term contracts to sell credits to buyers. Investors in forestry projects need to assess the volatility of carbon credit prices and consider whether to hedge future revenue by entering into an agreement with a buyer to sell future issuance of credits.

Most important, investors in forestry projects must determine whether the carbon credits generated will be eligible for sale in the compliance or voluntary carbon markets.

COMPLIANCE MARKETS

Carbon markets created and enforced by governments are called compliance markets because participants must keep below a cap or emissions threshold or risk paying a government-imposed fine. Compliance markets are also regulated by a government entity, which defines additionality and project eligibility. Compliance markets have the advantage of rule of law, ensuring that participants face penalties for noncompliance. As a result, the value of carbon credits in compliance markets tends to be significantly higher than in voluntary markets that lack government oversight. In 2021, the average price of credits in compliance markets was nearly $35,[10] while credits in the voluntary markets averaged less than $5.[11]

California's cap-and-trade program is the largest compliance carbon market in the United States, established as part of the state's plan to cost-effectively reduce greenhouse gas emissions by 80 percent from 1990 levels by 2050.[12] California has issued 233 million carbon credits with a value of $7 billion since market launch in 2013.[13] More than 80 percent of those credits were earned by forestry projects, providing market evidence that trees provide a low-cost, scalable climate solution.

However, establishing a compliance carbon market requires governments to enact legislation that regulates greenhouse gas emissions. Regrettably, the political will to do so is often lacking, leaving many project investors with only the voluntary carbon markets.

VOLUNTARY MARKETS

In the absence of government action, developers have created voluntary carbon credits to offset greenhouse gas emissions at low cost. Voluntary carbon credit projects are identical to compliance projects, except that the regulator and issuer of the credits is an independent organization instead of the government. Voluntary carbon markets are used by individuals and corporations to offset their greenhouse gas emissions and are growing quickly because of demand from companies with pledges to reduce greenhouse gas emissions.

For example, Delta Air Lines CEO Ed Bastian announced in 2020 that the airline would become fully carbon neutral.[14] However, airlines have few options for reducing greenhouse gas emissions, as purchasing more fuel-efficient engines or using biofuels are costly and of limited environmental benefit. In his announcement, Bastian stated. "I don't ever see a future where we'll eliminate jet fuel from our footprint." Carbon credits provide a solution. Delta budgeted $30 million in 2020 to purchase 13 million voluntary carbon credits, primarily from forestry projects, to offset the airline's emissions from burning fuel.[15]

Investors in forestry projects can generate revenue by selling carbon credits into voluntary markets. But the voluntary markets face risks in addition to those found in the compliance carbon markets, in the form of lower prices and weaker demand. In the United States, credits issued for the California compliance traded at a premium price above $31 in 2021[16] versus an average of $6 in the voluntary carbon markets.[17] And voluntary markets face greater reputational risk.

Voluntary carbon markets are more lightly regulated than compliance markets, lacking government oversight and the ability of governments to enforce fines on market participants for noncompliance. This creates a risk to investors of "greenwashing," or making a claim of climate benefits that turn out to be illusory. An investigation by Bloomberg of forestry projects developed by The Nature Conservancy reported it sold voluntary carbon credits from forests it did not intend to harvest, inferring the projects are not additional and calling into question credits sold to J.P. Morgan, BlackRock, and Disney.[18]

To address these challenges, former Bank of England governor Mark Carney and Standard Chartered Bank CEO Bill Winters formed the Taskforce on Scaling Voluntary Carbon Markets, an oversight body with the mission of ensuring carbon credits are valid. The taskforce is expected to establish market standards for the methodologies that determine project additionality, impermanence, and leakage, and the process for verification and issuance of credits. Agreement on global standards will improve credibility and reduce risk to investors in carbon credit projects, allowing for rapid growth—McKinsey forecasts the voluntary carbon markets will grow 100-fold by 2050.[19]

As the voluntary carbon markets grow to meet demand from companies to offset emissions, the opportunity to invest in forestry projects will expand, especially overseas.

TROPICAL FORESTS

Most deforestation is taking place in developing countries. Tropical countries lost 12 million hectares of forest cover, equivalent in area to all of Belgium, in just 1 year.[20] Unsurprisingly, this means 90 percent of opportunities for forestry carbon offset projects are in developing countries.[21]

This has prompted investors to pursue much larger forestry projects, the most ambitious of which is the Trillion Tree Initiative, launched at Davos in 2020 with backing and support by numerous countries and philanthropists including Marc Benioff, CEO of Salesforce.[22] The impact of such projects could be substantial. A research paper in *Science* estimated that 25 percent of atmospheric CO_2 could be sequestered by planting an additional billion hectares of forest cover.[23] But investors in these forestry projects will confront many challenges.

Tropical forests are in some of the most remote parts of the world. Monitoring millions of hectares of dense jungle is challenging, while enforcing protection from illegal logging and clearing of land is even more difficult. There is also the risk of monoculture on reforested lands when native trees are superseded by plantation forests.[24] Finally, there is a geographic limit to how much forest can be planted, capping the potential impact of the sector overall. Scientists with the IPCC warn "the large potential of afforestation will diminish over time, as forests saturate."[25]

The greatest risk to investing in forest conservation in developing countries may come from those countries' governments. Land in protected forests cannot be used for agriculture or for most other sources of economic development, diminishing its value. Governments in developing countries often face pressure from local citizens to make land available for development, especially in countries experiencing food shortages. While the revenue from carbon credits can be substantial, it may not be enough.

Protecting forests provides a global benefit, reducing the atmospheric concentration of CO_2. But at the national level, protecting forests is often unpopular. As President Bolsonaro of Brazil put it: "We understand the importance of the Amazon for the world—but the Amazon is ours."[26] For most countries, economic development takes priority over climate change.

For investors, protecting standing forests and planting new forests offer attractive but challenging opportunities. Beyond forestry, two other climate solutions, BECCS and regenerative agriculture, are attracting investor attention.

BIOENERGY CARBON CAPTURE AND STORAGE (BECCS)

Trees and other biomass can both sequester CO_2 and generate zero-emissions electricity using a technology called bioenergy carbon capture and storage (BECCS). BECCS uses biomass as a fuel source in a thermal generator, then captures and stores the resulting CO_2 emissions using the CCS technology described in chapter 8. BECCS is attractive because it has two positive climate attributes: biomass such as trees absorb and sequester CO_2 during years of growth, and then become a carbon-neutral source of electricity when burned. The U.S. National Academy of Sciences estimates that up to 5.2 billion tons of CO_2 could be sequestered globally by 2050 using BECCS.[27] The appeal of BECCS is obvious, as are the risks.

BECCS requires successful protection and growth of forests, with all the challenges described earlier in this chapter. But BECCS also requires technology to capture and store CO_2, which is costly and only at the pilot stage of development. The total cost of BECCS is difficult to forecast given the uncertainty of using carbon capture technologies at scale, with estimates ranging from $20 to $200 per ton of CO_2, dramatically higher than the cost of sequestration simply from planting trees.[28] In spite of the high cost, several utilities are experimenting with BECCS. Drax, a large utility in the United Kingdom with a plan to become a carbon-negative company, is using BECCS in two pilot projects to sequester emissions and generate electricity.[29] However, the sites are not expected to reach scale until 2030. For investors, BECCS offers a tantalizing opportunity to finance projects that sequester greenhouse gases and generate power, but that potential remains mostly theoretical until the cost of carbon capture technologies declines significantly and at scale.

REGENERATIVE AGRICULTURE

When trees and other plants die, the carbon they absorb through photosynthesis becomes part of the soil. The management of that soil determines how quickly the carbon is returned to the atmosphere. Planting of cover crops and low tillage on agricultural land, often referred to as *regenerative agriculture*, can reduce greenhouse gas emissions at very low cost, with the potential for increased crop yield and less soil erosion. The National Academy of Sciences estimates agricultural soil could sequester up to 5 percent of U.S. greenhouse gas emissions.[30]

Carbon markets provide a financial incentive, as investors in regenerative agriculture projects can be eligible to receive carbon credits. Like forestry, agriculture projects face the challenge of demonstrating additionality, ensuring that the farmer would not have changed practices anyway, but unlike forestry, it is easy to verify compliance on farms. The technologies to support regenerative agriculture are rapidly advancing, allowing for better data on soil health, management, and carbon sequestration. Most important, no-till farming can save farmers money and reduce greenhouse gas emissions at zero cost.[31]

Farmers tend to be conservative by nature, and many are skeptical of climate change and carbon markets, so its unsurprising that the market for regenerative agriculture carbon credits has been slow to develop.[32] That may soon change. Agricultural industry giant Cargill launched a voluntary program for the 2022 growing season that will pay farmers to capture more carbon in their soil, with a target of 10 million acres with regenerative agriculture by 2030.[33] In the near future, carbon may simply become another crop for American farmers.

FORESTRY, AGRICULTURE, AND CARBON MARKETS

Forestry projects and regenerative agriculture are nearly certain to expand simply because they offer an inexpensive, scalable climate solution, creating an attractive opportunity for investors. But these projects are surprisingly difficult to implement because generating revenue from carbon credits is complex and uncertain. Despite those challenges, carbon markets will play an important role in the era of climate change. McKinsey forecasts a global market for carbon credits worth up to $50 billion in 2030.[34]

FROM REAL ASSETS TO FINANCIAL ASSETS

Investing capital directly into real assets—including renewable energy, energy efficiency in buildings, forestry, and agriculture—requires sector-specific knowledge, deep pockets, and the expertise to negotiate complex financial agreements. It also requires know-how to generate and trade carbon credits. In most cases, only the largest institutional investors and companies have these capabilities. Individual investors and smaller institutions will instead use financial assets to gain exposure. The next section of this book explains the rewards and challenges of investing in financial assets in the era of climate change.

SECTION 5

Investing in Financial Assets

Financial assets span a range, from high-risk, high-return venture capital and private equity to public equities, funds, and fixed-income securities. Climate change will affect every one of those asset classes, introducing new risks and opportunities for businesses and investors. Larry Fink, founder and CEO of BlackRock, the world's largest manager of financial assets, predicts dramatic changes ahead: "It is my belief that the next 1,000 unicorns—companies that have a market valuation over a billion dollars—won't be a search engine, won't be a media company, they'll be businesses developing green hydrogen, green agriculture, green steel, and green cement."[1]

Larry Fink may or may not be proved right in his prediction, but every investor will benefit from understanding the risks and opportunities to their financial assets in the era of climate change.

"We bypass the animal, agriculture's greatest bottleneck."

FIGURE 18.1. Ethan Brown, founder and CEO of Beyond Meat. (Photographer: Tom Cooper, courtesy of Getty Images)

Chapter Eighteen

VENTURE CAPITAL

The macro trends of climate change—shifting public sentiment, new low-carbon technologies, and government incentives—create an opportunity for entrepreneurs to launch companies to mitigate the climate crisis. For entrepreneurs and venture investors, the potential to launch and finance innovative new businesses is unprecedented. As are the challenges.

A NOBLE WAY TO LOSE MONEY

Venture capitalists seek business sectors that have the potential for extraordinary growth and the outsized investment returns that come from backing innovative companies with the right product at the right time. In 2006, leading venture funds were convinced they had found the next great investment opportunity, a sector they dubbed "cleantech." Legendary Silicon Valley venture investor John Doerr, who made his reputation by investing early in Amazon, Google, and other successful start-ups, declared: "Green technologies—going green—is bigger than the Internet. It could be the biggest economic opportunity of the twenty-first century."[1]

Doerr was right, but his timing was wrong, by more than a decade. Following his lead, from 2006 to 2011, venture capitalists committed $25 billion to early-stage companies focused on solutions to climate change.[2] And promptly lost more than half of it.

The list of cleantech failures is long, but Solyndra is perhaps the most infamous. The company was founded to commercialize a new technology to produce cylindrical rather than flat solar panels, with purported gains in efficiency and improvements in cost. Unfortunately, by the time Solyndra commercialized its technology, the market had changed. Chinese competitors lowered prices, and declining raw material prices eroded Solyndra's relative cost position.[3] Ultimately, Solyndra filed for Chapter 11 bankruptcy protection, and venture capital investors lost roughly $1 billion.[4]

Solyndra was not the only high-profile venture-backed failure. Better Place was founded with a mission to end the demand for oil by an innovative solution for recharging electric vehicles using a network of battery swap stations that would allow drivers to quickly exchange their depleted battery for a fully charged one. In 2012, Better Place raised more than $800 million from top-tier venture investors and promptly collapsed less than 2 years later.[5]

What went wrong? Answering that question is crucial to understanding the challenges faced by investors in early-stage companies in the era of climate change. The unexpected failure of cleantech venture investing resulted in an exodus of investors after 2012 and low levels of financing for early-stage companies offering solutions to climate change.[6] As Joe Dear, chief investment officer of CalPERS, the largest pension fund in America, put it, "Our experience is that this has been a noble way to lose money."[7]

THE VALLEY OF DEATH

Venture capital investors in early-stage cleantech companies discovered multiple challenges to financial success:

1. *Physical products are slow to commercialize.* Companies with innovative solutions to climate change create mostly physical products; for example, solar panels to generate electricity. Physical products, by their very nature, experience long development cycles, requiring much testing and refining prior to commercialization. And that takes time. Which is a risk, as competitors and the market are moving targets, constantly improving their products. For example, Solyndra's product was compelling when the company was raising venture capital, but the time it took to develop and refine its solar technology left the company vulnerable to competitors with progressively lower-priced products. Physical products also suffer from long sales cycles, especially when customers are governments, utilities, or large corporations. These buyers want to see proof of concept, preferably market-tested

products, before committing to the purchase of new technologies. Unfortunately, proof of concept takes time, reducing the ability to pivot if the competitive landscape changes. Time works against entrepreneurs and venture investors, and physical products take time.

2. *Commodities are unattractive markets for new entrants.* The energy sector is the primary source of greenhouse gas emissions and the largest opportunity for investment. But energy, specifically electricity and fuel, are commodities, which means consumers select among alternative suppliers primarily on the basis of price. Price competition creates an advantage for large incumbent companies, making commodity markets inherently unattractive for new entrants and venture investors. A related problem is that low-carbon technologies such as renewable wind and solar are often selling into regulated markets that face political pressure to keep prices low. Regulated markets provide little incentive for innovation and risk-taking, favoring established incumbents and discouraging new entrants. Even worse, in the United States the demand for energy is inelastic. This means when innovation creates a lower-priced product, demand remains flat, squeezing margins. To succeed in an inelastic commodity market requires low-cost production, and that requires scale. Reaching scale requires huge sums of capital for construction of facilities and product distribution, creating yet another challenge.

3. *Capital intensity creates a "valley of death."* Many cleantech companies are only competitive at scale, often requiring $1 billion or more in risk capital. But venture capital investors rarely invest more than $100 million in early-stage companies. This financing gap is euphemistically known as the "valley of death," as many early-stage companies fail to cross it. A Brookings Institution study concluded that because of the funding gap, "VCs are reluctant to fund high-risk, capital-intensive ventures like offshore wind farms, biofuel refineries, and unproven solar cell technologies."[8]

4. *Government policies create uncertainty.* Innovative climate solutions often require government subsidies or regulatory support to initially compete with incumbent technologies. Unfortunately, government policies are inherently unstable, subject to the whims of politicians. The production tax credit for wind projects is a case in point, as congressional deadlock has allowed the credit to expire four times over two decades, creating turmoil in the wind sector. This political uncertainty creates a problem for entrepreneurs because venture capitalists will accept technology and commercialization risk but will avoid funding companies that face political and regulatory uncertainty, which lie outside of a company's control.

PIVOTING TO SUCCESS

Venture capitalists, to their credit, learned from the failures of cleantech investing in the early 2000s and pivoted to backing early-stage companies with business models designed to avoid the challenges facing the sector. Venture investors focused on companies with asset-light business models, shorter sales cycles, and products with brand appeal. The revised investment strategy worked. Venture returns on cleantech rebounded from negative 1.1 percent for the 2005–2009 investment period to +23.9 percent for investments made from 2014 to 2017.[9] Beyond Meat is an illustrative example of a climate solution with a strategy that led to great financial success.

Beyond Meat was founded in 2009 by Ethan Brown (figure 18.1) to create a meat substitute that would appeal to traditional consumers of beef. Brown understood that the incumbent product, beef, is inherently inefficient, requiring several pounds of grain for every pound of meat.[10] Beyond Meat's strategy is to "bypass the animal, agriculture's greatest bottleneck" with a product that is tastier, healthier, and more efficient to produce.[11]

Beef also plays an important role in climate change. Beef cattle alone are responsible for 6 percent of global greenhouse gas emissions because of methane emissions and grazing on deforested land.[12] Beyond Meat's products reduce emissions by more than 90 percent.[13]

Brown's strategy included branding to avoid commoditization and a targeted rollout strategy first to restaurants and then to full grocery retailers, to avoid competing head-on with incumbent meat companies at scale.[14] Importantly, Beyond Meat's scientists brought the first product to market within a year of raising venture capital, and then iterated to improve on the company's offerings through successive product launches.

Beyond Meat's strategy allowed it to develop innovative products with a modest amount of investment capital. The company raised $122 million in venture funding over 8 years, with Kleiner Perkins leading the first round.[15] It was a good bet. In 2019, Beyond Meat became a public company in the most successful initial public offering (IPO) of the year, surging 163 percent upon listing to value the company close to $4 billion.[16]

The most important lesson of Beyond Meat and other successful early-stage companies offering climate solutions is to create a better product, not just for the environment but for consumers. Research at the University of British Columbia on the value of environmental benefits offers an important warning for investors in early-stage climate solutions: "A frustrating paradox remains at the heart of green business: Few consumers who report positive attitudes toward eco-friendly products and services follow through with their wallets."[17]

The solution to this paradox is for entrepreneurs such as Ethan Brown to create products that are superior to the incumbents, offering better taste and potential health benefits. The *Wall Street Journal* summarized this point in an article on Beyond Meat: "It's not just good for the planet, it's also good for you."[18]

VENTURE INVESTING IN THE ERA OF CLIMATE CHANGE

Successful venture capital investments in climate solutions follow this playbook: identifying niches of the market where entrepreneurs can create a differentiated offering, then iterating and refining the offering as they scale, investing additional capital as required. With that strategy in mind, climate-focused venture investors are committing more than half of their capital to start-ups focused on mobility and transport, along with a substantial allocation to food and agricultural products.[19] Venture investors are betting electric vehicles can provide a cleaner product for the planet and a better-performing product for drivers and that consumers will continue to crave healthier, tastier foods.

Venture capital investors committed a record $37 billion to early-stage climate-focused companies in 2021, now called "climate tech," up more than 20-fold from a decade earlier.[20] When compared to all venture capital, investment in early-stage climate solution companies has experienced five times the growth rate of the overall venture market.[21] Thematic impact firms described in chapter 13, such as DBL and SJF, have raised increasingly larger venture funds, while leading traditional venture firms such as Kleiner Perkins have recommitted to the sector, raising $300 million for a second Green Growth Fund.[22] But some of the most aggressive investors in early-stage climate solutions are new players in the sector.

VENTURE FINANCING FROM NEW ENTRANTS

Large operating companies occasionally invest in early-stage businesses in their industry, a practice called corporate venture capital. Climate change, an existential threat to incumbent businesses in energy, transportation, cement, steel, and agriculture, is encouraging many companies to invest in climate solutions. Venture funding from corporations can bring strategic advantages, as the investors are also early customers and partners to many investee companies. For example, Energy Impact Partners, a consortium of 14 utilities, raised a $681 million venture fund to invest in starts-ups focused on next-generation technologies and business models.

Companies in the tech sector are also aggressively financing start-ups, despite facing relatively little risk from climate change. Microsoft created a $1 billion fund in 2020 with a specific focus on carbon-reduction technologies,[23] quickly followed by Amazon's $2 billion Climate Pledge Fund.[24] These climate mega-funds are designed to assist their tech company sponsors to meet pledges to reduce emissions to *net zero*, a topic covered in chapter 20.

Philanthropists are the most surprising, and possibly most important, new players in the venture capital space. Dubbed "philanthrocapitalists," these ultra-wealthy individuals and their foundations understand the catalytic potential for venture capital to finance solutions to climate change using a strategy called impact first investing. Breakthrough Energy Ventures, founded by Bill Gates and described in chapter 14, is emblematic. Unlike traditional venture investors, philanthrocapitalists can invest in opportunities that face great uncertainty, bridging the gap between basic science and commercial products. And these investors are patient, supporting early-stage companies through the long and perilous process of bringing physical products to market. It appears to be working, with Breakthrough Energy Ventures and other philanthrocapitalists financing successful investments in advanced batteries and other climate solutions. In 2021, Breakthrough Energy Ventures announced another $1 billion in funding, focused on tougher climate solutions such as green hydrogen and direct air capture technologies.[25]

VENTURE CAPITAL: RISK, RETURN, AND OPPORTUNITY

Investors considering venture capital financing of early-stage companies need to be acutely aware of the many challenges to commercialization of climate change solutions. But venture investors with a focused strategy can do very well. Kleiner Perkins recognized a 760-times return on its initial investment in Beyond Meat,[26] and thematic impact venture funds such as DBL and SJF have generated returns in the top quartile of all venture capital funds.[27]

For investors, venture capital is a high-risk, high-reward sector in which climate solutions will play an increasingly important role. The opportunity to invest in early-stage companies at the forefront of climate change is enticing even the most successful venture capitalists. Chris Sacca is well known in the venture world for his early and wildly profitable bets on Twitter, Instagram, Uber, and Stripe, before walking away from traditional venture investing in 2017. Four years later, Sacca returned to the sector, this time with a climate tech fund. Called Lowercarbon Capital, Sacca's new fund raised $800 million in "a few days" with the thesis that "fixing the planet is just good business."[28]

"Climate change is a one-way economic, fundamental shift.

FIGURE 19.1. Megan Starr, Global Head of Impact for the Carlyle Group (Source: REUTERS / Alamy Stock Photo)

Chapter Nineteen

PRIVATE EQUITY

Private equity investors have become active in the financing of climate solutions, motivated by limited partners seeking ESG-aligned funds, recognition of the risk of climate change to portfolio investments, and the potential to earn steady returns in renewable energy infrastructure. These trends are convincing private equity investors to shift capital away from the fossil fuel sector and towards low-carbon companies, attracted by the prospect of investing trillions of dollars in what Brookfield, one of the world's largest private equity firms, has called "the greatest commercial opportunity of our age."[1]

ESG AND THE TRANSITION TO NET ZERO

Asset owners committing capital to private equity fund managers are typically long-term investors such as public pension funds and insurance companies. Those asset owners often employ environmental, social, and governance (ESG) investing strategies, described in chapter 12. Asset owners are also asking private equity fund managers for commitments to net-zero emissions targets in their investment portfolios, forcing asset managers to evaluate climate risks and solutions.

To meet the demands of asset owners, private equity fund managers are employing two climate strategies: reducing emissions in current portfolio companies, and directing investment to companies with climate solutions.

LOWER EMISSIONS, HIGHER REWARDS

The commitment of portfolio companies to reduce emissions of greenhouse gases can generate long-term savings and enhance investor returns. Leading private equity firm Blackstone Group has committed to reducing carbon emissions by 15 percent within 3 years of buying any asset or company for its portfolio, primarily by increasing energy efficiency.[2] In doing so, Blackstone hopes to satisfy clients with ESG mandates while increasing portfolio company profits. Blackstone has significant experience with this strategy—portfolio company Hilton Hotels cut energy use by 22 percent over a decade, reducing emissions by 30 percent and saving more than $1 billion.[3]

Investing in energy efficiency improvements is a tried-and-true strategy for reducing emissions and saving money. Another, potentially more lucrative strategy is to invest directly in companies with zero-emission business models. Until recently, private equity has played a minor role in financing companies with climate change solutions, as most of those businesses were either too risky or too small. But rapid growth in several climate sectors, especially renewable energy, has created an attractive opportunity for private equity infrastructure funds.

RISE OF THE CLEAN INFRASTRUCTURE FUND

Infrastructure funds managed by private equity firms have a long history of financing traditional oil and gas projects and companies. Investing in renewable energy offers a natural transition as the energy sector shifts from fossil fuels to cleaner sources. Private equity investors are increasingly attracted to renewable energy solar and wind assets using proven technologies that deliver consistent long-term financial returns. Even better, the renewables sector has reached scale, requiring more than $50 billion a year in capital in the United States and more than $300 billion globally, more than large enough for even the most well-capitalized private equity funds.[4]

Unlike venture capital investors, private equity infrastructure funds avoid technology and development risk, acquiring projects when they are operational. Forecasts for rapid growth in renewable energy is prompting many private equity firms to launch clean energy investment funds. BlackRock was an early leader among large asset managers, raising its first renewable energy fund in 2011 and channeling $5.5 billion into investments in renewables since then.[5] Martin Torres, head of renewables at BlackRock, affirmed "it's now the single largest area of private-market investable opportunities for infrastructure globally."[6] Not to be outdone, Brookfield launched a $7.5 billion fund with a focus on renewable energy.[7] Dozens of other private equity firms have launched similar, smaller funds.

Investors in private equity infrastructure funds face the same risks and returns as investors placing capital directly into operating solar and wind projects, as described in chapter 4. For most investors, funds make more sense than direct investment in projects, as funds provide investors with project sourcing, due diligence expertise, and a diversified portfolio. In compensation for those services, fund investors pay fees, typically 1–1.5 percent in management fees and up to 20 percent in carried interest for renewable energy infrastructure funds.[8] For all but the most sophisticated investors, infrastructure funds are less risky than direct financing of solar and wind projects.

DIFFERENT FIRMS, DIFFERENT STRATEGIES

Private equity asset managers are committing to strategies for addressing climate change and capitalizing on investor demand for ESG funds. Some are incorporating climate change as a factor in their flagship funds, while others are establishing new funds targeted at climate as an investment opportunity. In the former category, Blackstone requires all portfolio companies to report on energy use, led by CEO Stephen Schwarzman's belief that "it ends up being good economics."[9] Other private equity asset managers are taking a different strategy on climate change by evaluating risk.

Carlyle Group, a private equity firm managing $260 billion, goes further than measuring energy use, factoring climate risks such as rising sea levels and extreme weather into its investment decisions. In 2020, Carlyle published its inaugural assessment of climate risks, becoming one of the first private equity firms to complete a risk assessment using the guidelines of the Task Force on Climate-Related Financial Disclosures (TCFD; described in chapter 10). Megan Starr of Carlyle explained the strategy: "It's the same private equity model, but you have to expand your aperture of expertise and data and analysis and include a lot more science."[10]

Carlyle's belief is that measuring and evaluating climate risks will create a more climate-resilient portfolio, leading to better risk-adjusted financial returns.

AN IMPORTANT, LIMITED ROLE

Private equity firm Blackstone is the world's largest owner of real estate, making its focus on improving energy efficiency and reducing greenhouse gases a meaningful contribution to addressing climate change.[11] Similarly, the launch of climate solutions funds by leading private equity firms such as Brookfield and Carlyle ensures private equity will play an important role financing clean energy infrastructure. But the private equity market is a fraction of the size of the much larger public markets. In the era of climate change, companies with climate solutions will ultimately source capital in the highly liquid public equity and fixed-income markets.

"We have grown from an average-sized utility by market capitalization 15 years ago to the largest utility company in the world today."

—JAMES ROBO, CEO, NEXTERA

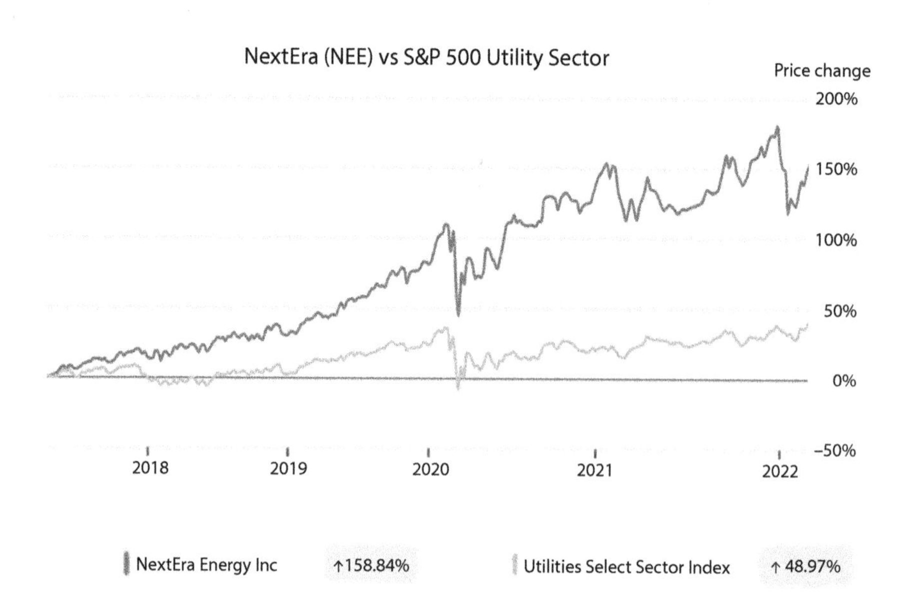

FIGURE 20.1. The graph illusrates investment returns for NextEra (NEE) versus the S&P 500 Utility Sector. (Source: Data from Google Finance, figure by the author)

Chapter Twenty

PUBLIC EQUITIES

Public equity markets provide investors easy access to companies that have climate solutions. Selecting which shares to buy is, however, anything but easy. Nearly every listed company has announced plans to address climate change, implementing strategies to reduce emissions while increasing profits. For investors, knowing which companies will succeed has become critical to investing in the public equity markets. Forecasting corporate winners and losers in the era of climate change begins with an understanding of how companies plan on reducing emissions.

THE TRANSITION TO NET ZERO

Climate scientists have established that keeping global temperatures from rising no more than 1.5°C will require a reduction in global greenhouse gas emissions to zero by 2050. This has prompted government leaders to announce plans to reduce country emissions to *net zero*, allowing for some emissions as long as they are offset with negative emissions using carbon capture (described in chapter 8). More than 100 countries have pledged to reach net-zero emissions by 2050, including the United Kingdom, Canada, and the European Union.[1] Even China has pledged to reach net zero by 2060. The United States has a target to reduce emissions by 50–52 percent by 2030 and to reach net zero no later than 2050.[2]

Public companies are also making net-zero pledges. American Airlines has a 2050 net-zero target, Verizon pledged net zero by 2040, and Apple has promised to reach net zero by 2030.[3] Microsoft went one step further, announcing: "While the world will need to reach net zero, those of us who can afford to move faster and go further should do so. By 2050 Microsoft will remove from the environment all the carbon the company has emitted either directly or by electrical consumption since it was founded in 1975."[4]

Meeting these pledges will require a transition to net-zero greenhouse gas emissions, a process that is fraught with uncertainty. For business leaders, the transition to net zero will be challenging, but it also creates an opportunity for competitive advantage. Investors need to understand how companies will be affected by the transition to net zero and consider the implications for investment risks and returns.

A SIMPLE CONCEPT, COMPLEX TO DO

The transition to net zero is a simple concept—scientists have established the need to reduce global emissions to zero by 2050, and companies have agreed to transition to net zero by reducing emissions and by offsetting emissions that cannot be otherwise eliminated with carbon credits. But the path to reaching net zero is made exceedingly complex by the following issues:

- *Scopes 1, 2, and 3*: A company's greenhouse gas emissions come from many sources, including manufacturing, use of electricity, even employee travel. Companies also source intermediary goods from suppliers, which generate emissions, and sell to consumers whose use of products may further contribute to climate change. Establishing the source of emissions is the first step to reducing them.

A company's greenhouse gas emissions are of three types, or *scopes*. Scope 1 is direct emissions from company-owned and controlled sources; for example, on-site manufacturing. Scope 2 is indirect emissions from consumption of energy. For most companies, this means emissions from the generation of electricity by the utility providing power. Scope 3 is all indirect emissions not included in scope 2, primarily emissions from upstream suppliers and downstream consumers. Knowing a company's scope 1, 2, and 3 emissions is an important first step for investors when evaluating net-zero emissions pledges. From there it becomes complicated.

• *Inconsistent targets*: Company pledges to reach net-zero emissions are remarkably inconsistent. Ford and GM made similar announcements to reduce emissions to net zero by 2050, but Ford includes scope 3 emissions in its pledge and GM does not.[5] Unsurprisingly, the emissions from manufacturing cars, scopes 1 and 2, are much lower than the emissions from driving cars, scope 3. On the other hand, GM pledged net zero in its manufacturing operations by 2040, with Ford targeting 2050. The inconsistencies in targets are not unique to the automobile industry, as company pledges are entirely voluntary and unregulated.

• *Double and triple counting*: Emissions are often counted by more than one company, resulting in double and even triple counting. For example, a utility that generates electricity by burning coal will have high scope 1 emissions, and the company that uses that electricity to manufacture a product will have high scope 2 emissions. If the utility converts its coal plant to renewable energy, its scope 1 emissions will decline to zero. But so will the scope 2 emissions of the manufacturing company that buys electricity. If both the utility and the manufacturer claim a reduction in emissions, as they will do to meet their net-zero targets, then reductions will be double counted, exaggerating progress in addressing climate change.

• *Multiple pathways*: Companies have multiple pathways to reach the same objective of net-zero emissions. Section 2 of this book introduced climate solutions that reduce greenhouse gas emissions, including renewables, energy storage, electric vehicles, and green hydrogen. Companies are selecting from among these climate solutions to optimize for their unique situations. For example, Duke Energy, the largest emitter of CO_2 in the United States,[6] has committed to net zero by transitioning to wind, solar, and storage.[7] Ford plans to power all of its manufacturing plants with renewable energy by 2035 and to invest in electric vehicles.[8] Baker Hughes is focusing on energy efficiency, green hydrogen, and carbon capture.[9] And Occidental Petroleum is aggressively investing in direct air capture.[10] These companies have all set the same target of reaching net-zero emissions by 2050 but have selected pathways with vastly different risks and hurdles.

• *Reductions versus removal*: Corporate net-zero pledges invariably include at least some use of emissions offsets or carbon removal, in addition to emissions reductions. Nearly every business can reduce emissions using climate solutions, but no business can eliminate them because some tasks, for example air travel, are likely to continue to emit greenhouse gases for many decades to come. The solution to these hard-to-abate emissions is carbon removal, introduced in chapter 8, in the form of sequestration in trees and other plants, capture and storage, or direct air capture technologies. In theory, reducing emissions and removing emissions should have the same beneficial result—fewer greenhouse gases in the atmosphere and less warming of the climate. In reality, there is much controversy.

The challenges of carbon removal and the use of carbon credits were covered in chapters 8 and 17, presenting companies with the unenviable choice of paying less for emissions reductions with reputational risk or paying more for carbon removal technologies that are unproven and costly. Forest sequestration is cheap yet unreliable, while carbon capture and storage technologies are an expensive climate solution, and direct air capture even more so. Every company with a net-zero pledge must plan on including some carbon removal along with emissions reductions, but the solutions for doing so present a host of challenges.

CREATING A COMPETITIVE ADVANTAGE

Despite the many complexities accompanying corporate net-zero pledges, the advantages can significantly outweigh the hurdles. Boston Consulting Group has studied the cost of emissions reductions for its corporate clients and found that the expense of reaching net zero by most companies is surprisingly low. For example, the additional cost of manufacturing an automobile with zero-carbon materials is just 2 percent, and for electronics the cost differential is less than 1 percent.[11] That is a small price for companies to pay given the trends affecting business in the era of climate change.

Physical changes from climate change—rising seas, drought, wildfires, and severe heat—pose a risk to companies that fail to prepare for them. *Technological innovation* to create climate solutions is upending industries from automobiles to meat. *Evolving social norms* are increasing consumer demand for low-carbon products and socially responsible companies. Recognizing these three trends, a fourth trend—*government regulations and incentives* at the federal, state, and local levels—are accelerating change. Companies that understand and stay ahead of these trends are likely to gain a competitive advantage over their rivals.

Business leaders clearly understand the advantages accruing to companies viewed as front-runners in the transition to net zero. In a survey of 250 senior executives, two-thirds of them agreed that "a company's net zero transition strategy and leadership is now a better predictor of future corporate success than its past financial performance."[12] There is another, even more important reason for a company to reduce emissions: a price on carbon, sooner or later.

A PRICE ON CARBON

Greenhouse gas emissions are a negative externality, as the consequences are suffered by future generations and not by today's polluters. Negative externalities are tackled with government regulations, by enacting a cap-and-trade system to limit emissions or imposing a carbon tax that forces polluters to pay for the future harm of emissions. In either case, the objective of regulators is to put a price on carbon that reflects the cost to society of reducing emissions. Once carbon emissions are priced, emitters act quickly to reduce them as it is in their best interests to do so.

Greenhouse gas emissions are, however, an unusual negative externality, as the impacts are global: emissions from anywhere affect everyone, everywhere. In an ideal world, greenhouse gases would be regulated internationally, thereby addressing the global dimension of this negative externality. Unfortunately, that is extremely difficult to do. Section 1 of this book described the challenges faced by governments attempting to regulate emissions and the failure to reach a binding international agreement to address climate change. But failure to put a global price on carbon does not mean it cannot be priced at the national or local level.

The European Union enacted the first significant regulation on greenhouse gas emissions, establishing a cap-and-trade market in 2005 that covers emissions from 11,000 companies.[13] The EU policy allows European companies to trade the right to emit carbon with each other, creating a price on carbon. The success of the EU market has been imitated in many other countries, most importantly China.[14] Other countries have implemented carbon taxes, including Canada and Sweden. As of 2022, nearly 22 percent of all greenhouse gas emissions globally are regulated by government caps or taxes, with carbon prices ranging from $1 to $137 per ton of CO_2.[15] The United States has not regulated greenhouse gas emissions with a cap or tax—not yet.

THE INEVITABLE POLICY RESPONSE

The UN's Principles for Responsible Investment is preparing companies for government regulation of greenhouse gas emissions, an initiative called the "inevitable policy response." The rationale is compelling: "As the realities of climate change become increasingly apparent, it is inevitable that governments will be forced to act more decisively than they have so far."[16]

Many U.S. companies are getting ahead of this outcome by lobbying for a price on carbon. Surprisingly, this includes large emitters of greenhouse gases such as ExxonMobil. After years of lobbying against climate regulations, ExxonMobil CEO Darren Woods in 2021 wrote "carbon pricing would send a clear signal through the market, creating incentives to reduce emissions."[17] For companies emitting greenhouse gases, a price on carbon would create an additional cost to doing business, but it would also bring certainty, allowing for development of climate solutions that require long-term planning and investment, such as carbon capture projects and green hydrogen.

CLEAR TRENDS, DIFFICULT CHOICES

The four major climate trends described in section 1 of this book are convincing companies to reduce greenhouse gas emissions, and a potential fifth driver, a price on carbon, would further accelerate the transition by businesses to net zero. Those trends are clear and inexorable. But knowing the direction of the market or an industry does not necessarily inform investors as to which companies will be winners. That often comes down to a battle between incumbent companies and disruptive innovators, the outcome of which determines leadership in each business sector.

INCUMBENTS VERSUS DISRUPTORS

Investors in public equities must decide between incumbent companies that currently dominate a sector or disruptive companies that have the potential to lead in the future. Selecting among these alternatives is difficult. Two examples of incumbents versus disruptors during the era of climate change, Tesla versus GM and NextEra versus SunEdison, provide investors with lessons on the advantages of each.

Tesla Versus GM

General Motors (GM) is the quintessential corporate incumbent, the largest automobile manufacturer in America since 1929.[18] GM sold 6.3 million vehicles in 2021, earning $10 billion on revenue of $127 billion.[19] In spite of its success, GM has been dramatically overtaken by Tesla in the public equity markets, with Tesla valued at approximately 10 times the value of GM in 2022. Understanding Tesla's success provides lessons that can be applied by investors to other sectors in which disruptive companies are overturning the status quo.

Chapter 5 described Tesla's path to launching the world's first successful electric vehicle company. Tesla made three important strategic decisions: iterating products starting with the high end of the market, focusing on product performance rather than environmental benefits, and taking advantage of declining costs in lithium-ion batteries. GM could have made the same decisions but did not. Why did the disruptor succeed in this case?

Incumbents have R&D budgets that dwarf the resources of disruptive firms, and incumbent firms such as GM can and do innovate. It is a myth that large incumbent firms are unable to invent new products and bring them to market. (GM built the first modern electric vehicle, the EV1, in 1996, nearly a decade ahead of Tesla.) The reality is that incumbents are poor at managing change. Stanford professor Bill Barnett points out that "it's not that big companies are bad at inventing; it's that they're bad at organizational shifts."[20]

Disruptive firms such as Tesla succeed by competing with companies in sectors in which organizational change is difficult. For example, many automobile companies are locked into agreements with car dealerships whose profits stem mainly from servicing and repairs. Introducing electric vehicles, which require little servicing, puts those profits at risk, disincentivizing dealers from supporting innovation in the auto sector. The lesson for investors is to seek disruptors in sectors where incumbent companies struggle to respond due to organizational challenges.

Tesla is famously well-known for its market success. But a much lesser-known company, NextEra, an industry incumbent, has been nearly as successful. The triumph of an incumbent also offers many lessons for investors.

NextEra Versus SunEdison

SunEdison was an early disruptor in renewable energy, developing solar and wind projects, initially in the United States and then internationally. SunEdison used innovative financing to support rapid organic growth and aggressive acquisitions to become the world's largest developer of renewable energy projects. Investors were enamored by the company's progress and the potential for solar and wind to replace fossil fuels, and SunEdison's stock soared, rising more than 2,000 percent from 2012 to 2015.[21] And then it collapsed. Worse-than-expected earnings and high financial leverage led to a liquidity crisis in the fall of 2015 and a desperate search for capital, culminating in a filing for bankruptcy in April 2016 and the loss of all shareholder capital.[22]

Like SunEdison, a utility called Florida Power & Light also recognized the commercial opportunity of solar and wind energy. Beginning in the early 2000s, the newly named NextEra methodically built a portfolio of wind and solar projects across the country, making the most of the company's competitive advantage: financing. NextEra's incumbent utility business is a monopoly in Florida, where it generates steady cash flow, giving the company the ability to fund its expansion into renewables with low-cost capital. Analysts at Morningstar concluded as much, writing "the utility business provided the sound balance sheet to finance renewables."[23] NextEra's strategy paid off for investors, creating $125 billion in shareholder value and a 1,082 percent return to investors over 15 years, dramatically outperforming the utilities sector index (figure 20.1).[24] In 2021, NextEra was the most valuable utility company in the world.[25]

The lesson for investors is to seek incumbents in sectors in which disruptors face barriers, including regulations, capital, or customer access, all of which create "strategic moats" for sector leaders.

Picking Winners

Investors selecting stocks in the era of climate change should determine which incumbents are at an advantage. Incumbents in the energy sector, such as NextEra, have several strategic moats because the sector is regulated, capital intensive, and produces a price-competitive commodity. In more nascent sectors, for example carbon capture, incumbents are well positioned to invest in early-stage companies or acquire companies with proven technologies and then commercialize them, as Occidental Petroleum is doing.

Conversely, disruptors are more successful in market sectors confronting new social trends, as Beyond Meat has done, and in sectors in which large incumbents face disincentives to organizational change. In most cases, disruptors succeed with products that offer an improvement on the status quo, often at a higher price. Elon Musk put it simply: "There is a lot of potential if you have a compelling product and people are willing to pay a premium for that."[26]

But disruption alone does not guarantee corporate success, even in sectors ripe for change. Analysis by investment advisory firm Research Affiliates highlights the challenge: "In a competitive industry, market disruption benefits society at large, not necessarily the disruptors, and disruptors are often disrupted themselves in due course."[27]

In the era of climate change, disruptive companies are benefiting society by offering climate solutions, but not all of them will reward their financial backers. Investors in public equities must differentiate between sectors that can offer a branded premium product, such as automobiles, and sectors offering commodity products, such as energy, and then select the companies best positioned for market dominance.

Investors in public equities should also be aware of two unusual public equity structures used to finance climate solutions, SPACs and YieldCos, and the advantages and challenges associated with each.

SPACs

Special purpose acquisition companies, or SPACs, are public companies established for the purpose of acquiring a private company. They are sometimes called "blank-check companies." The SPAC structure can be attractive to private growth-stage companies, offering access to public equity markets earlier and more quickly than the traditional initial public offering (IPO) process. For companies with climate solutions, SPACs are especially attractive when there is a need for vast amounts of capital to fund rapid growth. In 2020 alone, SPACs acquired climate solutions companies valued at $4 billion, including several electric vehicle (EV) manufacturers, an EV charging infrastructure company, and an energy storage technology company.[28]

SPACs provide investors with an opportunity to purchase equity in public companies that are difficult or impossible to access as private companies. For investors seeking to put capital into climate change solutions, SPACs greatly expand the breadth of potential investments. But investing in SPACs can be risky. Dave Kirkpatrick of growth equity firm SJF Ventures warns that "public markets can be fairly unforgiving" when revenues or profits are delayed.[29] The SPAC structure does not solve the challenges faced by growth-stage companies described earlier, other than to provide significant funding. By committing capital to SPACs that fail to generate meaningful revenues and profits, SPAC investors may find themselves repeating the mistakes of venture capital investors in the cleantech bubble that ended in 2008.

YIELDCOS

Like SPACs, yield companies (yieldcos) are public companies designed to purchase assets. Unlike SPACs, yieldcos focus on the relatively low-risk business of operating wind and solar projects, offering investors a modest but stable dividend. In many ways, yieldcos are the clean energy equivalent of a real estate investment trust (REIT), which acquires and holds a portfolio of real estate assets, distributing cash flows in the form of dividends to shareholders.

Yieldcos are created by moving operational solar and wind projects off the balance sheet of a development company and into a newly established company, then selling equity in the new company in an IPO. This is attractive to the development company as yieldcos trade at a higher multiple by holding de-risked operating assets with long-term power purchase agreements generating stable long-term cash flows. Yieldco investors receive a consistent dividend yield and expectation of dividend growth as new projects are developed and sold by the development company to the yieldco. NextEra established a yieldco in 2014, along with several other large renewable energy development companies.[30]

Unfortunately, yieldcos suffer from an inherent weakness in corporate governance. To grow dividends, yieldcos must continually purchase projects, creating a conflict of interest with the development company over the price of acquiring each new project. This weakness has occasionally been abused by the development company, most egregiously in the case of SunEdison, which unloaded low-grade assets onto its yieldco, "stacking the board with loyalists who would do its bidding" according to investor David Tepper.[31] SunEdison's bankruptcy put a halt to the launch of new yieldcos, and this investment structure is now limited to a small number of public equities.

Given the strong demand from investors for yieldcos, but the inherent weakness in the model, this investment structure has been mostly replaced by funds and ETFs, described in the next chapter.

BlackRock's U.S. Carbon Transition Readiness ETF was "the biggest launch in the ETF industry's three-decade history"

—BLOOMBERG

FIGURE 21.1. Solar panels being installed (Source: Wikimedia Commons)

Chapter Twenty-One

EQUITY FUNDS

Equity fund managers were slow to recognize the risk of climate change in their portfolios or to fully grasp the investment opportunity presented by the transition of the global economy in the era of climate change. That changed in 2020 with the Net Zero Asset Managers initiative, signed by 87 firms representing $37 trillion in assets under management.[1] Signatories committed to the goal of net-zero greenhouse gas emissions by 2050 and agreed to set interim targets at least every 5 years. By agreeing to a common goal and individual accountability, asset managers are hoping to reduce risks and profitably finance climate solutions, an opportunity that is too big and too important to miss. The world's three largest asset managers—BlackRock, Vanguard, and Fidelity—all participate in the Net Zero Asset Managers initiative.[2]

Climate-focused funds and ETFs (collectively "funds") fall into several categories, mirroring the investing strategies described in section 3 of this book. For investors with a focus on avoiding climate risk in their portfolios, *low-carbon funds* offer a solution. Investors wanting to avoid fossil fuel companies altogether can place capital in *fossil fuel–free funds* that shun oil, gas, and coal companies by use of active or passive investing strategies. Alternatively, investors seeking exposure to the companies best prepared for the transition to net zero can select from *climate transition funds* that incorporate environmental, social, and governance (ESG) analysis in the selection of portfolio holdings. And thematic investors with a focus on solving climate change can choose *climate solutions funds* with holdings in climate tech and renewable energy.

LOW-CARBON FUNDS

The world's largest asset managers are nearly unanimous in their assessment of the risk that climate change poses to investment returns. Vanguard's CEO believes that climate change "represents a long-term, material risk to our investors' portfolios,"[3] and Fidelity's chief executive declared "climate change poses one of, if not the most, significant risks to the long-term profitability and sustainability of companies."[4] Rounding out the top three investment firms, BlackRock stated: "We are firmly convinced that climate risk—physical and transition risk—presents one of the most significant systemic risks to the long-term value of our clients' investments."[5]

To address climate risk, asset managers are creating low-carbon funds designed to underweight companies with exposure to the climate risks described in chapter 10. Low-carbon funds can include fossil fuel companies but are structured with a total portfolio exposure to carbon that is lower than the overall market. In this way, low-carbon funds provide investors with downside protection in the event of a price on carbon or other transition risks.

State Street, in collaboration with Harvard Business School, researched portfolio strategies in both the United States and Europe that reduced carbon emissions while optimizing financial returns.[6] Their analysis evaluated six strategies for building long/short portfolios at the firm, industry, and sectoral level using carbon intensity to rank firms. The portfolios were rebalanced annually and controlled for changes in the price of oil over the period 2009–2018. This strategy delivered alpha of 2 percent annually over the period studied, providing investors with "significantly superior returns and continuous exposure to low carbon portfolios."

Low-carbon funds typically offer investors well-diversified portfolios. Unsurprisingly, low-carbon funds tend to overweight technology, consumer, health care, and financial services companies, having fewer industrial and energy companies than the overall market. For investors concerned primarily about risk, low-carbon funds offer an elegant solution. For investors with stronger feelings about climate change, fossil fuel–free funds might offer a better solution.

FOSSIL FUEL–FREE FUNDS

Chapter 11 introduced divestment, a strategy to avoid financing fossil fuel companies. Divestment is simple in theory but difficult for fund managers to implement as public companies often operate in multiple sectors with fossil fuel exposure. As a result, large asset management companies have struggled to offer funds that completely exclude all companies with exposure to fossil fuels, leaving the market to smaller, more focused firms such as Calvert, Parnassus, and Trillium. Those specialist firms offer actively managed equity funds with no fossil fuel holdings.[7]

Oxford University has pioneered a divestment strategy that allows active fund managers to engage with fossil fuel companies that are on a path toward lower CO_2 emissions while divesting from those that are unable or unwilling to do so. The *Oxford Martin Principles* encourages investors to support companies that are reducing greenhouse gas emissions to net zero by 2050—necessary to stabilize the climate and stay within the +1.5°C warming target of the Paris Accord—while divesting from companies that fail to meet milestone targets.[8] This sophisticated strategy is challenging to implement but has the potential to optimize investment returns while minimizing greenhouse gas emissions. Unsurprisingly, it is becoming an increasingly popular divestment strategy for active portfolio managers.

DIVESTMENT STRATEGIES FOR PASSIVE FUND INVESTORS

Passive investors seek to match a market index, which creates a challenge for divestment because most market indices include fossil fuel companies. To solve that problem, fund managers look to fossil-free market indices. Several ETFs have been launched that track the popular MSCI indices of stock markets in the United States, Europe, and emerging economies while removing companies with coal or oil and natural gas reserves.[9]

In 2015, State Street launched the SPDR S&P 500 Fossil Fuel Free ETF, building on the success of the SPDR S&P 500, the largest ETF in the U.S. equity markets. State Street's ETF tracks S&P's fossil fuel–free index, which excludes companies in the S&P 500 that hold reserves of thermal coal, crude oil, natural gas, or shale gas. Both ETFs exhibit similar financial returns, with the fossil-free index outperforming by 21 bp annually since its launch.[10] The SPDR S&P 500 Fossil Fuel Free ETF has grown assets under management to become the most popular listed divestment fund, offering investors a liquid security with zero exposure to fossil fuels while generating a slightly above-market return.[11] BlackRock, the world's largest manager of ETFs, announced in 2020 a plan to create up to 150 sustainable ETFs for passive investors.[12]

Sophisticated long-term passive investors can even try to outperform the market by selectively divesting fossil fuel companies at greatest risk from government regulation of greenhouse gas emissions. A research paper from Patrick Bolton of Columbia Business School found that this divestment strategy would allow investors to match market returns prior to governments acting on climate change and outperform the market when emissions regulations are enacted. With this passive divestment strategy, investors are in effect holding a free option on carbon that will result in financial outperformance when carbon dioxide emissions are priced by government regulation.[13] An estimated $50 billion in assets are managed using this strategy.[14]

CLIMATE TRANSITION FUNDS

Climate transition funds invest in public equities that demonstrate leadership in the transition to net-zero greenhouse gas emissions. These funds follow an ESG investing strategy whereby portfolio managers evaluate company net-zero transition plans in addition to more traditional fundamental analysis. The strategy of climate transition funds is to overweight companies best positioned to reach their net-zero targets ahead of their sectoral peers, creating a competitive advantage. Rapid growth in ESG as an attractive investing strategy, explained in chapter 12, has fostered significant investor interest in climate transition funds.

In 2021, BlackRock launched the US Carbon Transition Readiness ETF, an actively managed fund investing in large- and mid-cap U.S. equities that the portfolio manager believes are better positioned to benefit from a transition to a low-carbon economy. BlackRock's Carbon Transition Readiness ETF pulled in $1.3 billion on its first day of trading, making it the biggest launch in the ETF industry's three-decade history.[15]

Investors in climate transition funds stand to benefit from the transition to a net-zero global economy. Top holdings in BlackRock's US Carbon Transition Readiness ETF include Apple, Microsoft, and Amazon, companies with aggressive net-zero targets and detailed plans for reaching them.[16]

Investors in fossil-free funds, low-carbon funds, and climate transition funds have a reduced exposure to climate risk, which may lead to better financial returns versus market benchmarks. But the improvement in returns may be incremental, and the impact on solving climate change is modest. For investors seeking the potential for more substantial outperformance and greater impact mitigating climate change, climate solutions funds are a better fit.

CLIMATE SOLUTIONS FUNDS

Investing in companies with technologies and business models to address climate change—such as renewable energy, electric vehicles, and energy storage—offers the potential for significant financial rewards and impact. Companies in these sectors are experiencing rapid growth, as described in section 2 of this book, creating an ongoing need for capital. Direct investment in companies and projects is a viable strategy for sophisticated institutional investors, but most investors prefer diversified climate solutions funds and ETFs. Assets under management of climate solutions funds have expanded rapidly, especially in funds focused on climate tech and real assets.

Climate Tech

Funds investing in public companies that have climate solutions, also known as climate tech, offer investors exposure to sectors experiencing extraordinary growth. For example, the solar sector has grown at a compound annual rate of 42 percent in the past decade with similar progress expected in the years ahead,[17] the electric-vehicle sector is forecast to increase nearly 10 times in the next decade,[18] and the green hydrogen sector is projected to expand 13 times by 2025.[19] High growth rates can create attractive investment opportunities.

Many climate tech funds are highly specialized, focusing on a single technology because climate solutions are complex and require narrow sector expertise. Invesco's Solar ETF tracks a global index of solar companies, while First Trust's Wind Energy ETF invests exclusively in public companies in the wind power sector. Specialized funds provide investors with the opportunity to target specific climate solutions.

Other funds are more diversified. BlackRock's iShares Global Clean Energy ETF has garnered the largest assets under management by tracking an index of global equities across the clean energy sector.[20] Top holdings include wind turbine company Vestas, renewable energy utility NextEra, hydrogen technology company Plug Power, and climate technology companies like Enphase and First Solar.

Climate tech funds can generate very high returns. In 2020, iShares Global Clean Energy ETF gained an impressive 141 percent.[21] Unsurprisingly, climate tech funds are also highly volatile; the same ETF lost 24 percent in 2021.

Climate tech funds hold shares in many companies, offering investors greater diversification than investments directly in public equities, but returns on climate tech shares are highly correlated, creating an illusion of risk-reducing diversification with little actual downside protection. For investors, climate tech funds offer an exciting opportunity to profit from the next wave of climate solutions, while recognizing that the risks are high.

Real Assets

While climate tech funds offer investors high risks and high returns, real asset funds are at the opposite end of the risk-reward spectrum, offering low risk and modest returns. Section 4 of this book introduced investment in solar, wind, and storage projects, but only deep-pocketed investors with technical expertise have the resources and ability to commit capital directly to projects. The alternative for most investors is real asset funds, which aggregate portfolios of operating projects and distribute cash flow as gains. In many ways, real asset funds have the same attributes as YieldCos, but with better corporate governance.

Greenbacker Capital Management, a specialist fund manager, launched one of the first real asset renewable energy investment vehicles in 2014, aggregating solar and wind projects throughout North America. Unlike private equity renewables funds, Greenbacker is open to public investors. Generating steady annual investment returns between 4 percent and 10 percent, with low volatility in performance, Greenbacker's assets surpassed $1 billion in 2020.[22] Larger asset managers have launched similar real asset funds.

Carbon Credits

Climate solutions funds include some unusual asset classes, such as the opportunity to invest in rising carbon prices. The KFA Global Carbon ETF, launched in 2020, seeks to provide investors with a total return benchmarked to global carbon prices.[23] The increasing regulatory pressure on companies to reduce greenhouse gas emissions is adding to demand for carbon credits and allowances, especially in Europe where the price of EU Allowances increased 143 percent in 2021.[24] Investors expecting a tightening of carbon emissions regulations could see further gains in the KFA Global Carbon fund, along with portfolio diversification as carbon markets exhibit a low correlation with other asset classes.[25]

EQUITY RETURNS, EQUITY RISK

Asset managers are expanding the number of climate-focused equity funds and ETFs, given the underlying trends driving companies to address climate change and the demand from investors for financial products connected to the transition to a net-zero future. Equity funds offer investors the advantages of technical expertise, diversification, and access to new and growing sectors. But equity returns are volatile and risky. While climate change and the transition to net zero are a given, steady investment returns are not. For investors seeking less risk in the era of climate change, fixed-income products can offer an attractive alternative.

"We've issued more than $2.5 billion in green bonds for environmental projects. Why do we do all this? Because solving problems—creatively and elegantly—is at the very heart of what makes Apple, Apple."

FIGURE 22.1. Tim Cook, CEO, Apple (Source: Wikimedia Commons)

Chapter Twenty-Two

FIXED INCOME

The fixed-income markets are the optimal source of capital for financing climate solutions because the debt capital markets are 10 times as large as the equity markets, providing the low-cost capital needed to finance renewable energy, energy storage, hydrogen, and carbon capture.[1] The IPCC estimates a need for $2.4 trillion in annual investment until 2035 to achieve a 1.5°C path.[2] Fortunately, the global fixed-income markets, with average annual issuance of $21 trillion, can provide the liquidity to finance the net-zero transition required to avoid catastrophic climate change.[3]

Many climate change solutions lend themselves well to debt financing, requiring large up-front capital investment and generating predictable, long-term cash flows. Fixed-income investments in climate solutions have gained in popularity with issuers and investors, as evidenced by rapid growth in the aptly named green bond market.

GOING GREEN

The World Bank pioneered a bond in 2008 with a simple yet innovative design: "A 'plain vanilla' fixed income instrument in which the issuer commits to earmark the bond proceeds for projects with environmental or climate-related benefit."[4]

The World Bank's innovation was called a green bond. Green bonds are similar to conventional bonds, with one key difference: the bond issuer promises to use the proceeds to fund one or more green projects. The definition of "green" was initially determined by the issuer, leading to some absurd bonds, including from Chinese companies for "clean coal" that was neither clean nor green.[5] This led to accusations of greenwashing, pretending to protect the environment while not actually doing so. To address this reputational risk, leading bond underwriters created the Green Bond Principles, an agreed upon set of standards for the issuance of green bonds and the evaluation of environmental impact. In the case of the World Bank, green bond proceeds have been used to finance renewable energy, energy efficiency, and clean transportation projects, and the World Bank issues an annual report on the impact created by those investments.[6]

Issuer Benefits

Issuers were initially attracted to green bonds to tap socially minded investors. As interest in environmental, social, and governance (ESG) investing grew, issuers found an expanding pool of buyers seeking debt products with environmental benefits. The responsibility for ensuring that bond proceeds were properly allocated to green projects created an administrative and reporting burden for issuers, but these tasks were more than made up for by the benefits. As the green bond market expanded, issuers realized several advantages.

Research on green bonds finds them more closely held than conventional bonds, a potentially more stable investor base.[7] Investors may even be willing to pay more for green bonds, creating a "greenium" over conventional bonds,[8] although the research on this is mixed.[9] Some, issuers of green bonds are touting the savings—Verizon issued a $1 billion green bond that the treasurer estimates saved the company $1.4 million every year in interest expense.

Green bonds might also benefit shareholders. Surprisingly, research by Caroline Flammer at Columbia University found that the share price of issuers rose around the announcement of a green bond offering, especially for first-time issuers, suggesting that equity investors also appreciate the merits of green bonds.[10]

Selling Out in an Hour

Green bonds are popular with fixed-income investors as they offer all the benefits of a conventional bond with the knowledge that the proceeds will be used to address climate change or other environmental challenges. The credit rating on green bonds is identical to that of conventional bonds from the same issuer, with similar yield and liquidity. What makes green bonds attractive is their simplicity—investors can finance climate solutions without any project expertise and without taking any project risk. The green bond investor has only to assess the credit quality of the issuer, not of the underlying projects, as the investor is repaid even if the green projects fail.

Unsurprisingly, investors have been snapping up green bond issues. Citibank underwrote a World Bank green bond that sold out in an hour, prompting Citi's managing director to proclaim, "We've reached the Holy Grail, which is the bond market."[11] While hyperbolic, the comment reflects the importance of accessing debt capital to finance solutions to climate change.

Criticisms

Green bonds are not without critics. Even with the Green Bond Principles, risk of greenwashing remains, as the standards are self-regulated and there are no penalties for violations. More important, projects financed with green bond proceeds do not require additionality, meaning the projects may have been financed anyway. Without additionality, green bond issuance does not promote incremental investment in climate solutions beyond what was already planned. Criticisms of green bonds are valid given that some issuers simply repackaged existing loans into green bonds, but the market has adapted and grown.

Innovation and Growth

The green bond market is red hot. Annual issuance rose from less than $50 billion in 2015 to more than $500 billion in 2021, including many U.S. corporations and even the New York Transit Authority.[12] Proceeds from green bond issues have funded thousands of projects aimed at reducing greenhouse gas emissions. To cite just one example, Apple (figure 22.1) announced that projects funded through issuance of green bonds will reduce greenhouse gas emissions equivalent to removing nearly 200,000 cars from the road.[13] Analysts predict further growth in the green bond market, driven by demand from climate-focused fixed-income investors attracted to the simplicity of the green bond structure and by companies keen on issuing debt that is less costly and reputation enhancing.[14]

Along with market growth, green bond underwriters and issuers are introducing innovations to the product structure. Enel, a large Italian utility, issued a "sustainability-linked bond" with a coupon that steps-up 25 basis points if the company misses targets for reducing greenhouse gas emissions.[15] Other issuers are experimenting with "transition bonds" to finance the replacement of fossil fuels with lower-emissions alternatives.[16] Asset managers have even launched several green bond funds and ETFs, offering investors diversified portfolios overseen by professional managers.

Green bonds are an imperfect climate-financing product subject to criticism, but by tapping the highly liquid and low-cost debt capital markets, green bonds have become an important investment product in the era of climate change. Fixed-income products financing climate solutions can also be found in the market for asset-backed securities.

SECURITIZING SOLAR

Solar projects generate stable, long-term cash flows, making them attractive to fixed-income investors. Unfortunately, residential and commercial-scale solar projects are simply too small to directly access the fixed-income markets. Securitization, the process by which a pool of assets is placed in a financial vehicle to create an asset-backed security, has proven to be a key enabler for deploying capital into solar leases and loans.

Solar Leases

American homeowners in the early 2000s were reluctant to install solar systems that cost an average of $40,000 with uncertain performance and payback.[17] The introduction of solar leasing changed that. Solar leasing companies such as SolarCity offered to install residential solar systems in return for a 25-year lease, requiring no up-front investment by the homeowner. The solar lease required payment only for electricity generated, transferring project performance risk from the homeowner to the solar leasing company. The appeal of solar leases—no money down and no product risk—overcame homeowner reluctance, and solar installations soared. By 2014, solar leasing accounted for 72 percent of all new residential installations in the United States.[18]

Solar leasing companies signed up thousands of residential customers, creating a need for low-cost capital to finance installations. SolarCity undertook the first solar lease–backed securitization in 2013, bundling a $54 million portfolio of residential solar assets into a BBB+ rated asset-backed security.[19] Investors were attracted to a new fixed-income asset with long duration, attractive yields, and a history of stable payments. Subsequent growth in the residential solar leasing market led to more than $2 billion in annual issuance of solar lease–backed securities.[20]

Solar Loans

Over time, homeowners became familiar with solar, prompting a move away from solar leasing and toward solar loans that finance the outright purchase of residential solar systems. Companies such as Mosaic offered financing to solar installers, who then offered loans to homeowners along with panel installation. By 2018, solar loans had overtaken solar leases as the most popular financing product in the U.S. residential market.[21]

Like solar leases, solar loans can be aggregated and securitized. Mosaic issued the first asset-backed solar loan security in 2017, issuing a total of $1.1 billion over the next several years.[22] Solar loan asset-backed securities (ABS) issues are composed of homeowner loans with average FICO scores above 700, resulting in investment-grade credit ratings.[23] Investors are attracted by this fixed-income product's diversification across thousands of residential solar systems, stable cash flows, and low credit risk.

Securitization Catches On

Asset-backed securities are an attractive financial asset for fixed-income investors and an increasingly important source of capital for climate solutions. Securitization is certain to grow as investors become familiar with the advantages of diversified pools of renewable energy assets with long operating lives and stable cash flows. Even better, securitization has been a key factor in bringing down the cost of solar systems for U.S. homeowners, further increasing demand for residential solar systems and accelerating the transition from fossil fuels to renewables.[24]

While green bonds and solar securitization offer new investment opportunities, an entirely different fixed-income market, municipal bonds, is increasingly exposed to climate change risks. Analysts and underwriters are starting to take notice.

MUNICIPAL BONDS DISCOVER CLIMATE RISK

The U.S. municipal bond (or "muni") market has outstanding issuance of $3.9 trillion. Historically, muni bonds have suffered very low defaults, an almost negligible 113 out of more than 50,000 issuers.[25] By comparison, corporate defaults are nearly 100 times as frequent. But climate change is putting the exceptional repayment record of muni bonds at risk, as heat stress, rising seas, wildfires, and natural disasters impair municipal finances.

Surprisingly, most muni bond issuers provide little to no disclosure of climate risk. In contrast, credit rating agencies are already considering those risks and evaluating how cities are preparing to address them. Moody's, which has ratings on more than 500,000 government bond issues,[26] acquired a data company with the expertise to assess physical climate risks for more than 3,000 counties.[27] Putting that know-how to work, Moody's AA+ rating of a muni bond issued by Miami Beach was partly due to the agency's assessment of climate risk: "In our view, the city maintains among the most robust plans attempting to address [climate change] risks that we've reviewed for US local governments."[28]

Underwriters are also starting to ask questions. J.P. Morgan's head of public finance disclosed that discussion of climate risk with muni issuers is very much a part of their due diligence.[29] And Mellon Bank's head of municipal bonds believes it is only a matter of time before investors include climate risk and resiliency plans in their assessment of issuers.[30] The challenge for investors is that climate risks vary significantly by city and by type of muni bond.

General Obligation Versus Revenue Bonds

Municipal bonds are of two types, general obligation bonds and revenue bonds, each of which is exposed to climate change in different ways.

General obligation (GO) bonds are considered low risk by investors because municipalities have the power of taxation. In theory, municipalities can increase tax revenue in the event of a climate-related shortfall. But that depends on the nature of the climate risk. Acute events, such as a hurricane, can be addressed with taxes and federal aid. Chronic climate events, such as rising seas and frequent flooding, might lead to a downward spiral of declining housing prices, lower real estate taxes, and municipal deficits, putting GO muni bond payments at risk.

Revenue bonds are backed by cash flows from a specific tax or user fee. Disruption in revenue from a climate event, for example a severe storm or wildfire, can harm bondholders. The devastating 2018 California Camp Fire destroyed 18,800 structures and resulted in 88 fatalities, with an intensity owing in part to climate change–induced drought. Credit ratings on revenue bonds affected by the Camp Fire were cut, and bond prices tumbled—the California Infrastructure and Economic Development Bank's local revenue bonds declined 11 percent in 1 month.[31]

Pricing Risk

Muni investors should understand the specific climate risks faced by a community and the potential impact on bond prices. Some muni bonds are already trading at a slight discount because of climate-related risks, with a research paper finding that risk of sea-level rise has reduced muni bond prices by 2–5 percent in flood-prone areas.[32] Investors might be inclined to rely on the federal government to cover climate-related losses, but an increase in severe storms, wildfires, and flooding may weaken support for costly bailouts. Instead, investors should focus on municipalities with a resiliency plan designed to withstand climate-related events and protect bondholders.

Muni Bonds Go Green

American cities at risk from a changing climate are issuing green bonds, using the financing to invest in climate solutions and resilient infrastructure. The first muni green bond was issued in 2013 by the Commonwealth of Massachusetts to finance a program to reduce energy consumption at 700 sites, saving the state on energy costs and reducing greenhouse gas emissions. The $100 million issue was oversubscribed, attracting several new investors who had never previously purchased the state's bonds.[33]

Muni green bonds are popular with investors for the same reasons that make corporate green bonds popular—the simplicity and creditworthiness of a conventional muni bond combined with the added benefit of knowing that the proceeds will be used to address climate change. U.S. muni green bond issuance rose from $100 million in 2013 to $20 billion in 2020, and analysts predict further growth in the years ahead.[34] Eyeing growth in the market, Franklin Templeton, one of the largest muni bond fund managers in the country, launched a muni green bond fund for investors seeking to align long-term investment goals with their environmental values.[35]

FIXED-INCOME INVESTING IN THE ERA OF CLIMATE CHANGE

Equity markets receive most of the attention from analysts and the media, but the reality is that the bulk of the financing for climate solutions will be provided by the fixed-income markets, where capital is both low cost and available at the scale required for the transition to a net-zero economy. In the era of climate change, fixed-income investors must evaluate the risks facing bond issuers and should consider buying green bonds and asset-backed securities that finance solar. These fixed-income assets offer potentially higher risk-adjusted financial returns, along with the psychic return of knowing that the investor's capital will be used to address climate change.

SECTION 6

The Investor's Dilemma

Investors understand that climate change will affect nearly every sector of the global economy. But climate change is playing out over decades. Will investors react ahead of events that may not occur in their lifetimes? What are best practices for investing in the era of climate change? Most important, will it matter?

"We created an entirely new environment to which our brain is not perfectly adapted."

—DANIEL GILBERT, HARVARD UNIVERSITY

FIGURE 23.1. *The Thinker*, Auguste Rodin (Source: Wikimedia Commons)

Chapter Twenty-Three

THE INVESTOR'S DILEMMA

The era of climate change is upon us. The key question, frankly the only question that matters, is whether humanity will avoid *catastrophic* climate change? Optimists point to climate solutions that are technically and commercially feasible. Conversely, pessimists remind us that humans have never in recorded history addressed a global challenge with a multi-decadal impact such as climate change. Even immediate threats are poorly managed, as evidenced by the uncoordinated international response to the COVID-19 pandemic. Long-term challenges such as climate change are much harder to solve because of the limitations of the human mind.

Humanity is hindered by biology, as the human brain did not develop to solve problems for next year let alone next century. The Harvard psychology professor Daniel Gilbert explains how this became a problem: "We're very good at clear and present danger, like every mammal is. That's why we've survived as long as we have. . . . The problem is that our environment has changed so rapidly because we've got this great big brain so we could navigate our ancestral environment, and lo and behold, what did we do? We created an entirely new environment to which our brain is not perfectly adapted."[1]

Human evolution has not prepared people to make decisions with long-term implications. Even worse, individuals fail to act when faced with uncertainty. The Nobel Prize–winning economist Daniel Kahneman demonstrated through his research that the human brain responds most decisively to issues it is certain about. Unfortunately, climate change is fraught with uncertainty, as humans have never faced it before. The global impact of climate change is highly predictable, but the local impact on a community or individual is much less certain, reducing the willingness to take action.

Adding to the challenge, people underestimate climate change because it is a nonlinear problem, meaning the rate of change in emissions and global temperature is increasing at an accelerating rate. Humans understand and act on linear developments but are poor at judging nonlinear trends. Climate change, which increases slowly at first and then accelerates over decades, is challenging to comprehend because most people extrapolate current trends at a linear rate, vastly underestimating both the speed and the impact of global warming.

Elke Weber summed up the problem of solving climate change while a professor of management and psychology at Columbia Business School. "In a way, it's unfair to expect people, homo sapiens, to do this kind of monitoring, to do this kind of decision making, because we're not wired for that."[2]

The investor's dilemma? The human brain cautions us to wait for more information and certainty before investing in climate change, even though investors know that climate change is coming. This failure to act has a high cost, which will only increase over time.

THE IMPORTANCE OF TIME

Most people understand the need to reduce greenhouse gas emissions if humanity is to avoid catastrophic climate change. What is less well understood is why the timing of emissions reductions is so important. Understanding why time is of the essence requires a brief explanation of the difference between *flows* and *stocks* of greenhouse gases.

Greenhouse Gas Flows

Flows represent the greenhouse gases added to the atmosphere over a period of time, usually measured annually. For example, global flows of CO_2 in 2021 totaled 33 billion tons.[3] Other greenhouse gas emission flows include methane (CH_4) and nitrous oxide (N_2O), but CO_2 is by far the largest. Scientists have determined that avoiding catastrophic climate change, a temperature increase of more than 1.5°C–2°C, will require a reduction in net CO_2 *flows* to zero by 2050 or soon thereafter.

The commercialization of climate solutions described earlier in this book—renewable solar and wind power, electric vehicles, energy storage, green hydrogen, and carbon removal—will dramatically lower annual emissions, or flows, of CO_2. This is a reason for optimism in the fight against climate change. But time is running out because the *stock* of CO2 is already dangerously high.

Greenhouse Gas Stocks

Stock represents the total sum of CO_2 accumulated in the atmosphere. To calculate the stock, instruments are used to measure the concentration of CO_2 in the atmosphere in parts per million (ppm). This measurement is critical because the greenhouse effect that warms the planet is caused by the stock of CO_2 in the atmosphere, not from annual flows. CO_2 remains in the atmosphere for up to 1,000 years, which means that the current stock of CO_2 will continue to warm the planet, even if no additional CO_2 is emitted.[4]

Scientists have determined that a concentration of CO_2 of 450 ppm is likely to warm the planet by 1.5°C–2°C, which is considered the maximum warming possible without serious risk of disruption to global economies and human welfare.[5] Atmospheric concentrations above 450 ppm create uncertainties in feedback loops, where warming creates changes to the planet that accelerate further warming. The scientific consensus from more than 1,000 scientists in 80 countries is unambiguous in predicting that continued emissions of greenhouse gases increase the likelihood of "severe, pervasive and irreversible impacts"[6] and in recommending that the concentration of CO_2 in the atmosphere be kept as close as possible to a maximum of 450 ppm.

NO TIME TO LOSE

The challenge in keeping within this target is that the concentration of CO_2 reached 420 ppm in 2022, and annual emissions or flows increase the stock by 2–3 ppm every year.[7] At "business as usual" rates, CO_2 concentrations will reach 450 ppm by 2035 and close to 500 ppm by 2050, well above the maximum threshold advised by scientists.

The commercial climate solutions described in this book can result in a significant decline in annual CO_2 emissions if implemented at scale. Compared to the "business as usual" scenario, climate solutions can reduce annual emissions by approximately 75 percent.[8] Carbon removal for the remaining hard-to-abate sectors can bring the global economy to net-zero emissions by 2050 or shortly thereafter.[9]

But the timing of the decarbonization process dramatically affects the cost.

THE COST OF DELAY

A delay in reducing greenhouse gas emissions will give rise to a greater accumulation of CO_2 in the atmosphere, accelerating warming and increasing the risk of catastrophic climate change. The longer the delay in reducing CO_2 emissions, the greater the cost, as every additional ton of emissions adds to the stock already in the atmosphere, remaining there for centuries.

Climate scientists have modeled thousands of emission reduction scenarios to understand the pathways for limiting global warming to 1.5°C–2°C. Figure 23.2 displays four different emissions pathways modeled by scientists

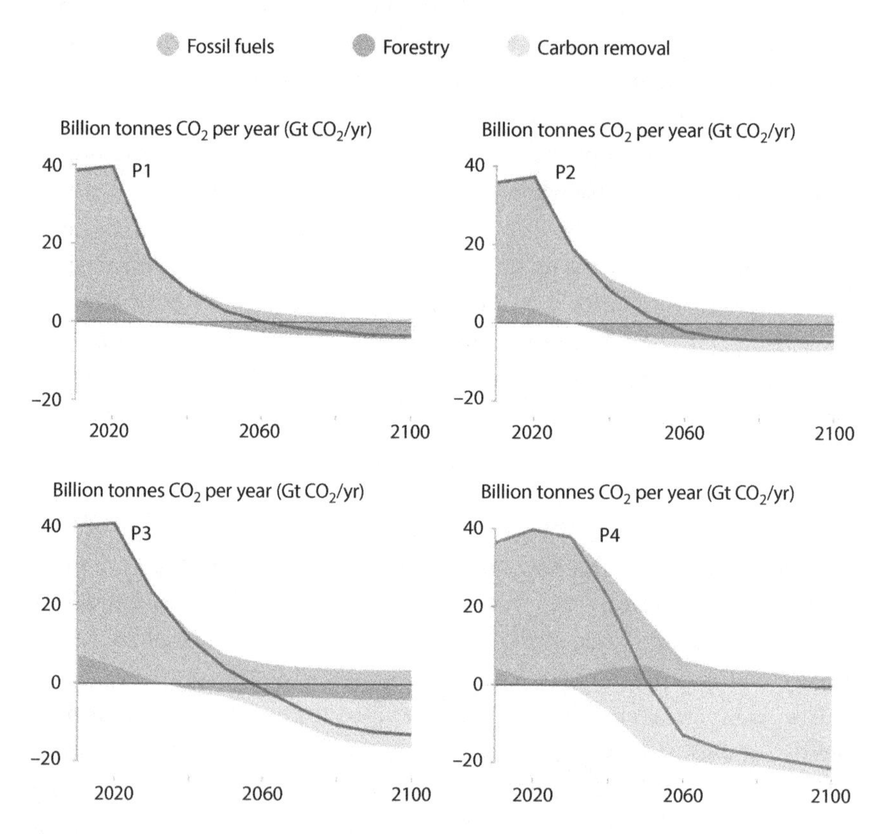

FIGURE 23.2. Model pathways limiting warming to +1.5°C (Source: IPCC)

for limiting warming to +1.5°C and avoiding catastrophic climate change.[10] The paths are labeled P1 to P4, with P1 representing a scenario of rapid decarbonization starting immediately. In the P1 scenario, global emissions decline rapidly to net zero with only a very modest amount of carbon removal, in the form of inexpensive forestry. P2 is a slower, less ambitious pathway and requires both forestry and other forms of carbon removal. Scenarios P3 and P4 demonstrate that a greater delay in reducing emissions must be counterbalanced with a greater reliance on carbon removal. As explained in chapter 8, carbon removal, especially direct air capture, is very costly.

The four emissions pathways to net zero described above all succeed in avoiding catastrophic climate change, but at very different costs. P1 is an ambitious path requiring no costly carbon capture. At the other extreme, P4 is a path of procrastination in which global emissions remain high for many years before dropping sharply with the use of costly carbon capture solutions. Professor Gary Yohe of Wesleyan University describes the cost of delay in reducing emissions more simply: "the longer you wait, the more expensive it gets."[11]

THE INVESTOR'S DILEMMA

Climate change is a challenge that no human has previously faced, let alone solved: a multi-decadal problem changing at a nonlinear rate with great uncertainty. Taking immediate action on climate change runs counter to everything the human brain is wired for. Moving quickly, however, is critical to keeping the cost of climate mitigation manageable. For investors, the dilemma is to overcome the human hesitation to act, making rational investment decisions that are timely without being rash. Knowing best practices makes that easier.

"Investors should 'climate proof' their investments by mid-century, in-line with the recommendations of climate scientists."

FIGURE 24.1. *Under the Wave off Kanagawa*, Katsushika Hokusai, Japanese woodblock print, ca. 1830 (Source: Wikimedia Commons)

Chapter Twenty-Four

BEST PRACTICES

Investors are entering a period of systemic change with few guideposts and an uncertain future. Despite the scientific and economic knowledge on climate change, much remains unknown. In the years ahead, there may be periods in which investors' assets are buffeted by the physical effects of a warming planet and by the political, technological, and social shocks of climate change, much as the mariners of nineteenth-century Japan were tossed about in a raging sea.

Navigating climate change will require investors to ignore the tendency to focus on near-term events while understanding how shifts will occur in an accelerating, nonlinear fashion. It will be complicated and hard. But the investor's task can be made easier by following a few basic principles, beginning with strategy.

ALIGN STRATEGY WITH OBJECTIVES

Every investor has a perspective on taking risk that will determine the investor's return objectives. Investors should select from among the investing strategies described in section 3, choosing the strategy that most closely reflects their financial and climate objectives.

Investors focused on minimizing risk should follow a climate risk mitigation strategy, reducing exposure to real assets that face physical risks and financial assets that face transition risks. Alternatively, investors focused on maximizing return should follow a thematic investing strategy, selecting securities with the greatest upside potential in the era of climate change.

Investors with the objective of mitigating climate change should follow a divestment strategy if their assets are modest and an impact first strategy if they are wealthy. And all investors should consider environmental, social, and governance (ESG) factors when evaluating assets, as this strategy can contribute to sound investment analysis.

Having selected an investment strategy, investors need to stick with it.

TAKE THE LONG VIEW

Timing a dramatic change in asset prices is exceptionally hard to do and rarely profitable. Most investors should take the long view, approaching climate change by selecting a climate investing strategy and then evaluating how their portfolios can reach net zero by 2050. To put it another way, investors should "climate proof" their investments by mid-century, in line with the recommendations of climate scientists. For many investors, 2050 will feel distant, but at less than 30 years it is within the time frame of pension funds, mortgages, and real assets. It is well within the time frame of most investors' retirement plans.

Having taken the long view on climate change, it is critical for investors to stay the course. Bill Gates offers a useful reminder: "We always overestimate the change that will occur in the next two years and underestimate the change that will occur in the next ten."[1] Investors have an unfortunate habit of overtrading, moving out of positions earlier than they should. Maintaining a long-term perspective is necessary to reap the benefits of a climate-focused investing strategy.

Taking the long view does not ensure all investments in climate solutions will be successful. But the odds of success can be improved by following five best practices specific to investing in the era of climate change.

BEST PRACTICES

1. *Beware of greenwashing, intentional or not.* Climate change has motivated business leaders, many of whom are well intentioned, to seek solutions. That is a noble thing. Unfortunately, good intentions alone rarely translate into business success. To address climate change, business models must be in alignment with climate solutions, not in conflict with them. The classic example is British Petroleum (BP), which in the early 2000s rebranded with

the tagline “Beyond Petroleum.” But the company continued to invest heavily in fossil fuels, committing 93 percent of capital to oil and gas projects, which spurred a backlash among nongovernmental organizations (NGOs) and consumers.[2] BP quietly dropped the new brand a few years later, a costly mistake by the company and for investors.

In 2020, BP announced plans to realign its business model to reduce emissions, by cutting oil and gas output 40 percent and growing renewable energy generation 20-fold.[3] The company is also investing heavily in green hydrogen and carbon capture. This time, the equity markets reacted warmly to BP's announcement, which more clearly aligned its business model with its climate mission, and the shares closed up 7 percent on the news.[4]

The lesson for investors is to beware of company greenwashing, which never creates long-term value, and commit capital only to companies with climate plans that align with core business models.

2. *The trend is your friend.* Investment opportunities in climate solutions often arise from trends in unrelated sectors. The rapid growth in consumer electronics gave rise to the demand for lithium-ion batteries, which were then used to create electric vehicles. The market for consumer electronics, and related demand for batteries to power them, was an obvious trend, and yet it was ignored for many years by the incumbent automobile companies. Looking ahead, the rapid growth and declining costs in renewable solar and wind power are a trend that will create an opportunity to develop green hydrogen. Similarly, the trend in company pledges to reduce greenhouse gas emissions will generate demand for carbon removal projects, creating opportunity in forestry, agriculture, and direct air capture.

Investors should be aware of the trends driven by or related to climate change and understand the impact across multiple sectors. Trends that are multi-year, or in some cases multi-decadal, will provide a powerful tailwind for business growth.

3. *Avoid businesses anticipating a change in human behavior.* Many a business plan has been written on the assumption that consumers are willing to change behavior and “do the right thing” for the environment. Unfortunately, few are willing to do so, especially if it requires prioritizing future concerns over current needs, what psychologists call *present bias*.[5] Businesses offering climate solutions are much more likely to succeed by offering products that require little or no change in consumer behavior.

For example, Tesla created the first successful electric vehicle company by offering a product that looks and drives like a traditional automobile, only better. Similarly, Beyond Meat intentionally designed its products to taste like real meat and to be available to consumers in the meat aisle at supermarkets. Ethan Brown, the CEO of Beyond Meat, summarized this strategy: "I have a strong commitment to meeting people where they are."[6]

The objective of Tesla, Beyond Meat, and other successful companies with climate solutions is not to fool the consumer—buyers know they are getting a different product—but to avoid the high hurdles to changing behavior. Investors should avoid businesses relying on a change in human behavior unless the customer niche is very focused, the change is minor, or the consumer benefits significantly outweigh the barriers to change.

4. *It's better to be early than late.* Investors are aware the climate is slowly changing, and there is an expectation that assets exposed to climate risk will gradually lose value. That could be a mistake. Economist Hyman Minsky described how markets suddenly and collectively reprice assets, referred to as a "Minsky moment." Asset prices are based on forward-looking estimates of future cash flows. If opinions on asset viability suddenly change, investors can be caught out, leading to feedback loops decimating asset prices. Regulators have warned of a "climate Minsky moment" brought on by a reassessment of climate risks that causes jump-to-distress pricing of exposed assets.[7] Investors should not wait for the market to reprice assets. It is advantageous to commit early and trade out of assets facing climate risk.

5. *All the normal investing rules still apply.* Climate change introduces new risks and new opportunities; it does not change the basic rules of investing. In the short term, asset prices are almost certain to be overvalued or undervalued, as occurs in all business sectors experiencing rapid transformation. But this does not alter the fact that long-run shareholder value is a function of current and future cash flows. Investors are advised to maintain investing discipline during the era of climate change.

Selecting a long-term investment objective and following best practices will position investors to earn attractive risk-adjusted returns in the years ahead, which will be good for investors. Will it be good for the planet? Will the global transition to climate solutions occur quickly enough to avoid catastrophic climate change? That answer depends, more than anything, on investors.

"Our planet is a lonely speck in the great enveloping cosmic dark. In our obscurity, in all this vastness, there is no hint that help will come from elsewhere to save us from ourselves."

—CARL SAGAN

FIGURE 25.1. Earth as seen from space (Source: Wikimedia Commons)

Chapter Twenty-Five

WHY INVESTMENT MATTERS

The American astronomer Carl Sagan had a gift for communicating astrophysics to a lay audience. By explaining the mysteries of outer space, Sagan forced people to recognize an important fact—we are on our own (figure 25.1). Human activity is rapidly warming the planet, and only humans can change that trajectory.

Record-breaking hurricanes, heat, and flooding are occurring with mind-numbing frequency in the United States and globally. But that is nothing compared to what lies ahead. Without a rapid and sustained decrease in emissions, the world will begin to experience significantly worse effects within decades, with the damage only increasing over time. The scientific consensus is that business as usual is "highly likely" to increase global temperatures more than 2°C, and possibly near 6°C, by 2100.[1] This will lead to widespread disruption of ecosystems and the global economy, along with the likely displacement of more than 100 million people.[2] The last time the earth was this warm was more than 3 million years ago, long before humans roamed the planet.[3]

"Business as usual" means continued use of our current infrastructure and technologies: burning fossil fuels for energy and transportation, emitting greenhouse gases from industry, deforesting land, and emissions-intensive agriculture practices. Emissions occur because the infrastructure of the global economy, created during the Industrial Revolution and built with trillions of dollars of investment capital, requires it. The only realistic solution to climate change is to reinvest in a global economy that does not emit greenhouse gases. Investment capital allows us to, as Carl Sagan wrote, "save us from ourselves."

Before considering the case for investment to create a low-carbon global economy, it is worth reviewing four alternative solutions for avoiding catastrophic climate change:

1. Consume Less

A survey of more than 10,000 consumers in 29 countries found half saying they could happily live without most of the items they own, relying on the sharing economy to meet many of their needs.[4] The rapid growth in peer-to-peer sharing of cars, homes, and even pets is reducing overall demand for some goods and services. Even better, some consumers are intentionally reducing consumption of emissions-intensive goods such as beef and air travel as their contribution in the fight against climate change. Unfortunately, their actions are unlikely to materially affect climate change.

The challenge with focusing on individual consumption is that it does not address most greenhouse gas emissions. Professor Michael Mann of Penn State explains why: "Focusing on individual choices around air travel and beef consumption heightens the risk of losing sight of the gorilla in the room: civilization's reliance on fossil fuels for energy and transport overall, which accounts for roughly two-thirds of global carbon emissions. We need systemic changes that will reduce everyone's carbon footprint."[5]

To make matters worse, studies find that people who make personal sacrifices to address climate change are *less* likely to support carbon taxes or other government regulations, suggesting that lowering individual consumption can erode support for climate policies.[6] This occurs because individuals forgoing personal consumption develop a false sense of the progress being made on climate change.

To be clear, reducing consumption of emissions-intensive goods and services is critical to reducing the use of energy and other inputs to address climate change. Energy efficiency is also one of the lowest-cost climate solutions. But individual actions to consume less will not be nearly enough to avoid catastrophic climate change, as global emissions must fall to *zero*, an objective that is impossible to reach using a lower-consumption strategy.

2. Abandon Economic Growth

Nearly all economists equate growth in GDP with success, ignoring the historic correlation between economic growth and greenhouse gas emissions. As an alternative, a few economists are now calling for an end to economic growth, or "degrowth," as a solution to climate change.[7] Unfortunately, degrowth suffers from two significant challenges.

First, developing countries are unlikely to cooperate, believing economic growth is their right as it has been for the more developed countries. The United States and Europe have benefited from the Industrial Revolution and the extraordinary economic growth that followed it, while polluting the atmosphere with greenhouse gases. To put that in context, the United States and Europe combined are responsible for nearly half of *all* greenhouse gas emissions emitted since 1751 despite having only 11 percent of the global population.[8] The other 89 percent of the people on the planet are highly unlikely to support a degrowth strategy that would limit them from attaining economic prosperity.

The second challenge is that even if countries were to follow a degrowth strategy, the outcome would only reduce emissions, not eliminate them, and at high economic cost. In 2020, global CO_2 emissions declined by 5.8 percent because of the COVID-19 pandemic, offering a real-life example of degrowth.[9] But the United Nations estimates that the global economy must cut emissions by 7.6 percent *every year* for a decade to keep warming below +1.5°C. Given the extraordinary hardship that accompanied the global economic downturn in 2020, it is unimaginable that most people would be willing to suffer an even greater economic decline, every year, for the foreseeable future.

3. Control Population Growth

Greenhouse gas emissions are a direct result of human activity, and some experts have suggested that the solution to climate change is simply to have fewer humans.[10] Setting aside the morality of limiting population growth or the practicality of doing so, a declining population would not avoid catastrophic climate change. The problem is timing. As one researcher put it: "Cutting the number of people on the planet will take hundreds of years. Emissions reduction needs to start now."[11]

A lower population would eventually reduce the flow of greenhouse gas emissions. But it would do little to reduce the stock of emissions. Controlling population growth would simply result in fewer people living on a much hotter planet.

Furthermore, it turns out that having children can affect adult behavior on climate change in a positive way. A research study found that children, especially girls, are effective at increasing their parents' level of concern about climate change.[12] The impact was strongest on conservative parents who tend to be less worried about global warming.

4. Adapt

Humans are the most adaptable of all species, and an argument can be made for adapting to climate change instead of trying to mitigate it. The Netherlands has demonstrated that it is possible to build barriers against rising seas at a manageable cost. Rising temperatures will impair crop yields in temperate and tropical countries, but arable land is expanding in northern regions, making Russia the world's largest exporter of wheat as a warmer climate lengthens the growing season.[13] Singapore, which lies almost directly on the equator, shows that air conditioning can allow for comfort and economic success, proving that hotter weather need not affect productivity or living conditions.

The problem with adaptation is that, as the economist Herbert Stein once quipped, "if it can't go on forever it will stop." Which applies to climate change. Eventually, humans must stop emitting greenhouse gases, as every additional ton of CO_2 and other greenhouse gases adds to the existing stock, making the planet hotter. At some point, rising temperatures will create dangerous feedback loops that accelerate emissions and speed up catastrophic warming.

Adaptation is a viable strategy in the short term, perhaps even through the end of the century. But business as usual will eventually warm the planet beyond the human capacity to adapt. Even worse, at that point it will be too late to do much about it. Adaptation is a false hope, a temporary benefit that will eventually lead to a very hot and dangerous outcome.

In fact, all four of the above solutions will *contribute* to addressing climate change. Less consumption, a reduced emphasis on GDP growth, and slower population growth are necessary but insufficient solutions. And some adaptation will be necessary, as the climate will continue to change even with a rapid reduction in emissions. That leaves only one realistic path for avoiding catastrophic climate change: aggressive investment in climate solutions.

THE CASE FOR INVESTMENT

Greenhouse gas emissions must reach net zero by 2050 or shortly thereafter. It is extraordinarily important that decarbonization of the global economy succeeds, as there will be no second chances to reduce emissions once the concentration of CO_2 and other greenhouse gases surpasses critical levels. The window for avoiding catastrophic climate change is already small and closing. Therefore, it is important to consider three key issues: How much capital is needed, where can it be sourced, and will investment be effective at avoiding catastrophic climate change?

How Much Capital Is Needed?

The Global Financial Markets Association, whose members represent the world's leading financial institutions, commissioned consulting firm BCG to determine how much capital is required to decarbonize the global economy and limit warming to 1.5°C–2°C. The sum came to $3–$5 trillion per year, or a total $100–$150 trillion by 2050.[14] Estimates from other analysts range from $1 trillion to $6.9 trillion per year, roughly in line with BCG's numbers.[15] Current investments in climate solutions total approximately $600 billion per year, implying a 5–8 times increase in capital needed for decarbonization.

Is It Available?

The short answer is yes, the required equity capital is available. BCG's analysis estimates that 65 percent of investment will use loans or bonds and the remaining 35 percent will use equity.[16] That equates to $2–$3.5 trillion per year in fixed income, and $1–$1.5 trillion per year in equity.

The fixed-income markets are very large and liquid, with $128 trillion outstanding and annual issuance over $20 trillion, suggesting that the additional capital needed to address climate change is readily available.[17] The equity markets are less liquid. Global equity markets are very large, totaling $86 trillion, but annual issuance of new equity is relatively infrequent, approximately $1 trillion per year.[18] The private equity and venture markets invest a further $0.5 trillion.[19] Raising an additional $1–$1.5 trillion per year in equity capital for climate change solutions is highly ambitious but theoretically feasible.

Will It Be Effective?

The prior sections in this book explained how greenhouse gas emissions can be reduced rapidly and cost-effectively by investing in climate solutions. Renewable solar and wind power, energy storage, and electric vehicles can eliminate more than half of all emissions. Green hydrogen can remove another 10–20 percent of emissions from transport and industry, and the remaining 15–25 percent can be abated using carbon removal technologies. *In theory*, there is a path for decarbonization to net zero by 2050. Which raises a further important question: Is there any evidence that investing in the climate solutions described in this book will be effective? Fortunately, some countries have already experienced early success.

Decarbonization with Growth

Germany's greenhouse gas emissions declined more than 40 percent from 1990 to 2020.[20] Remarkably, this was accomplished alongside a growing economy. During the same period, German real GDP grew by 46 percent.[21] In other words, Germany both decarbonized its economy and raised German prosperity.

Germany's decarbonization was accomplished by a dramatic expansion of renewable energy, named the *Energiewende*. Wind and solar power went from near zero to 45 percent of the country's total electricity generation, overtaking fossil fuels in 2020.[22] Germany's transition to renewable power required a massive investment in renewables, resulting in a hefty surcharge on consumer utility bills. But costs are declining rapidly. Looking ahead, German investment of 2.3 trillion euros is forecast to reduce greenhouse gas emissions by 95 percent by 2050, with a positive effect on the economy.[23] The country's most important business association conducted a study of the costs and benefits of the Energiewende and concluded "industrial companies will benefit from ambitious climate protection."[24]

Germany is not unique. A study of decarbonization found that 35 countries successfully decoupled economic growth from greenhouse gas emissions during the period from 2000 to 2014.[25] Among major economies, the

United Kingdom cut emissions by 24 percent while growing real GDP by 27 percent, an impressive accomplishment in a little over a decade.

To the surprise of many, the United States provides another example of decarbonization, albeit at a slower pace than Germany or the United Kingdom. In the 15 years to 2019, U.S. real GDP grew 27 percent and emissions declined 14 percent.[26] U.S. emissions declined because of replacement of coal-fired plants with investment in lower-carbon natural gas and zero-carbon wind and solar, along with greater energy efficiency in factories and homes, and more efficient vehicles.[27]

These results provide early evidence that countries can decarbonize and maintain economic growth. But as investors know from the warning label required by the SEC on every fund prospectus, "past performance does not necessarily predict future results."[28] Looking ahead, will the United States and the world accelerate the decoupling of the economy from greenhouse gas emissions and meet the science-based target of net-zero emissions by 2050?

PREDICTING FUTURE RESULTS

The National Academy of Sciences studied this question with a focus on the United States, gathering input from experts at more than a dozen leading universities. The study concluded that decarbonization of the United States is technically feasible, and a target of 2050 is realistic if the process starts immediately.[29] This will require $2.1 trillion in incremental investment by 2030, but savings on lower electricity costs will more than compensate for the up-front commitment.

The International Energy Agency (IEA) considered the question from a global perspective and reached nearly the same conclusion, finding a path to net-zero emissions that is feasible but requires enormous investment. To give a sense of the scale, consider that the IEA estimates the path to net zero will require installing what is currently the world's largest solar project every day for the next 10 years.[30] IEA executive director Fatih Birol described it this way: "The scale and speed of the efforts demanded by this critical and formidable goal—our best chance of tackling climate change and limiting global warming to 1.5 °C—make this perhaps the greatest challenge humankind has ever faced."[31]

Fortunately, the synergies among the climate solutions described in chapter 9 will further accelerate the path to net zero. The declining cost of wind and solar power, alongside the rapidly falling cost of energy storage, will accelerate the adoption of electric vehicles, further driving demand for renewables. Cheap renewable power will allow for massive development of green hydrogen and will underpin carbon capture technologies. Individually, the climate solutions will contribute to the reduction of greenhouse gases; together, they create a virtuous cycle that will accelerate the reduction of emissions and avoid catastrophic climate change. For investors, the synergies among the climate solutions will provide many opportunities to finance the greatest reinvention of the global economy since the Industrial Revolution.

What Are the Remaining Challenges?

The path to net zero is relatively clear, but there are several challenges to success. The greatest challenge is the absence of a price on carbon in most countries. Placing a price on carbon with an emissions cap or a carbon tax is essential to providing an incentive for all businesses to quickly invest in climate solutions and reduce emissions. Similarly, government subsidies remain critical for rapid implementation of climate solutions, especially with respect to green hydrogen and direct air capture, both of which will remain costly and commercially uncompetitive until further investment brings down costs.

In the financial markets, standardized and audited reporting on emissions and reporting of material climate risks is necessary to provide investors with reliable data for directing investment toward companies on a path to lower emissions and away from companies contributing to or at risk from climate change. Regulators need to ensure that reporting is reliable and consistent, and that companies treat climate data with the same seriousness of purpose as financial information.

Finally, perhaps the greatest challenge to reaching net zero is the inability by countries to cooperate. Climate change is a global problem, and climate solutions such as direct air capture will only be effective with cooperation on measurement and compensation for emissions reductions. Unfortunately, the rise of nationalism is making international cooperation ever more challenging.

Despite these challenges, the rapidly improving economics of key climate solutions means a significant reduction in carbon intensity is within reach, and net zero by 2050 is possible. But the decades ahead will be filled with extraordinary disruption, requiring tremendous commitment by government and the private sector. What does the future hold for investors and for the climate?

"We know exactly what's causing climate change. We can absolutely 1) avoid the worst and 2) build a better world in the process."

—KATE MARVEL, COLUMBIA UNIVERSITY

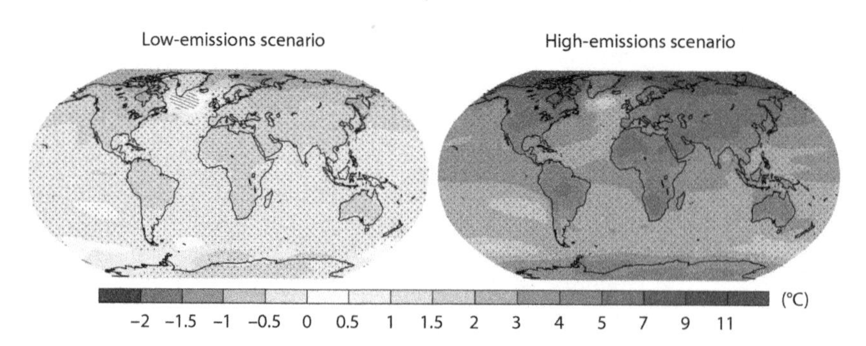

FIGURE 26.1. Projected change in average surface temperature between 2000 and 2100 (Source: IPCC)

Chapter Twenty-Six

THE FUTURE

The planet is warming, and many changes are coming that will be highly disruptive at best and catastrophic at worst. For investors, the next 30 years will be a time of increasingly rapid change, just as the past 30 years was, but for a completely different reason.

CREATIVE DESTRUCTION: 1990–2020

The Austrian economist Joseph Schumpeter described capitalism as a process of "creative destruction" in which technological innovation by new companies drives older firms out of business. Schumpeter believed this to be "the essential fact about capitalism."[1] Investors have proven him right, financing innovative companies that created enormous shareholder value while simultaneously destroying businesses that failed to compete. Schumpeter developed his economic theories in the 1940s, but it was a half-century later that his concept of creative destruction reached its zenith, with the advent of the digital age.

In 1998, Kodak's market capitalization was $26 billion, and Apple's was $1 billion.[2] Within 20 years, Kodak was forced to declare bankruptcy, and Apple was the most valuable company in the world. Film, the heart of Kodak's business, was replaced by digital photography and Apple's iPhone. Technology and the process of creative destruction destroyed one American icon and created another.

But this was not just about one successful company or one sector of the economy. Digital innovations affected every business in nearly every way, dramatically changing the value of financial assets. By 2020, the five most valuable companies in America were in technology.[3] Climate change is unleashing Schumpeter's creative destruction once again, and as with the digital revolution, great wealth will be made, and lost, in this process.

CREATIVE DESTRUCTION: 2020–2050

Predicting the growth of climate solutions such as renewable energy and electric vehicles is relatively easy. Predicting which companies will dominate those market sectors is much harder. As it was for selecting market winners in the past.

Investors in the 1990s understood that digital technologies would be an increasingly important factor in business, yet it was hard to know which companies to finance. Netscape was the leading Internet company in 1996 with nearly 80 percent of the browser market. Five years later, the company was nearly worthless. Google was another company with an Internet browser, but with a very different outcome for investors. Every investor wants to believe that he or she would have avoided Netscape and purchased shares in Google.

The era of climate change will force investors to make similar investment decisions, assessing the risks and opportunities in a rapidly changing environment. *The Economist* summed it up neatly: "Like the internet, decarbonization will lead to structural change in the global economy. Capital will have to flow towards cleaner technologies. The process will create winners and losers."[4]

Climate change will affect investors in a process of creative destruction that will last for at least the next three decades. Investments will be affected by climate change, and investors will affect climate change. The key question, perhaps the only question that really matters, is whether investors will have enough of an impact.

Will Humanity Avoid Catastrophic Climate Change?

There are many reasons for doubt. The relentless growth in greenhouse gas emissions and the inability of governments to reach a binding international climate agreement to lower emissions despite nearly 30 years of negotiations offer little hope for the future. Humanity is further burdened by a biology that struggles with challenges such as climate change. But all is not lost, as the writer Matthew King makes clear: "It's important to remember one thing, however. It's true that no other species has evolved to create such a large-scale problem—but no other species has evolved with such an extraordinary capacity to solve it, either."[5]

Humans have the solutions to avoid catastrophic climate change, but it is unclear whether humanity will implement those solutions in time. The greatest challenge by far is the inconsistency of government support. This creates uncertainty, delaying investment, and losing precious time. The good news is that the emissions curve finally appears to be bending downward because of the trends introduced in section 1, offering the first glimpse of a low-carbon future. Unfortunately, we may not get there in time.

WE ARE OUR OWN WORST ENEMY

Doubters, on both sides of the climate debate, make the path forward significantly harder. Climate denialists have slowed the implementation of sensible government support for climate solutions by sowing uncertainty and confusion. The denialists are losing influence as the science of climate change becomes more widely understood, and as the physical effects become increasingly apparent, but much time has been lost in the process.

At the other extreme, climate defeatists are ignorant of the advance of climate solutions and the potential for human ingenuity. By resigning themselves to catastrophe, the defeatists remove the desire to address catastrophic climate change until it becomes too late, creating a self-fulfilling outcome. Unfortunately, defeatists are becoming more influential as the climate warms.

The vast range of opinions and debates over climate change is wasting the most precious resource of all: time.

MOMENTUM, AT LAST

Time is tight, but the window has not yet closed for avoiding catastrophic climate change. The momentum that exists in low-carbon technologies, changing social norms, and government support represent a turning point in the global emissions of greenhouse gases, putting rescue within reach. Columbia University climate scientist Kate Marvel neatly summed up the situation when she tweeted: "As a climate scientist, I'd like you to know: I don't have hope. I have something better: certainty. We know exactly what's causing climate change. We can absolutely 1) avoid the worst and 2) build a better world in the process."[6]

Without doubt, this process will be slower than it should be due to international rivalries, inadequate coordination among nations, and pushback from businesses losing out in the transition to a low-carbon future. But humanity can avoid catastrophic climate change because it now has the solutions to do so. The delay in reducing emissions will necessitate the aggressive use of costly carbon removal solutions to reach net zero, making the path to a low-carbon future slow and costly rather than quick and cheap. But there is a path forward.

The Future

Scientists have discovered the causes of global warming and have made clear what must be done if humanity is to avoid a catastrophe. Engineers have invented climate solutions. Entrepreneurs and business leaders have created commercial applications. Politicians are reacting to changing social norms by providing more government support. Momentum is building for a rapid implementation of climate solutions at scale. The transition to a global low-carbon future has begun, providing investors in the era of climate change with the opportunity and challenge of a lifetime. Investors financed the extraordinarily successful Industrial and Agricultural Revolutions of the past, now investors have an even more important role: financing the world's future.

ACKNOWLEDGMENTS

I teach nearly 500 MBA students each year, and from every one of them I learn something. For that I am deeply grateful. From among those students, special thanks to my summer research assistant Sean Fleming, and to Siwol Chang, Monica Cho, Angelica Crispino, Brian McNamara, Caitlin Wischermann, and Shradha Mani for background research that contributed to this book.

The faculty at Columbia University have been enormously supportive, especially Columbia Business School Dean Costis Maglaras and Professor Geoff Heal, and the extraordinary team at the Tamer Center under the direction of Sandra Navalli. The leadership of Columbia's Climate School, Deans Alex Halliday, Jason Bordoff, Maureen Raymo, and Ruth DeFries, provided me with invaluable insights on climate science and policy, and informed me of the pressing need to avoid catastrophic climate change.

The entrepreneurs working on climate solutions, and the investors backing them, provided inspiration. There are too many to name, but special mention to the Community Energy team of Brent Alderfer, Eric Blank, and Brent Beerley, the Mosaic team of Billy Parish and Dan Rosen, and the investment teams at SJF Ventures, headed by Dave Kirkpatrick, and at Greenbacker headed by Ben Baker.

Myles Thompson and Brian Smith of Columbia University Press, thank you for your advice and support. It is a pleasure working with you.

Eric, your detailed review and comments on the draft manuscript made it immeasurably better. I am very lucky to have a sibling in a position of leadership in the climate sector with whom I can share ideas.

Naomi, this book would never have been written were it not for your support of my transition from a career in business to one in academia. You told me it was the best decision I would ever make, and you were right.

NOTES

PREFACE

1. Jennifer Marlon, Peter Howe, Matto Mildenberger, et al., "Yale Climate Opinion Maps 2020," Yale Program on Climate Change Communication, September 2, 2020, https://climatecommunication.yale.edu/visualizations-data/ycom-us/.
2. "Future Climate Changes, Risks and Impacts," in IPCC 2014 *Synthesis Report*, accessed February 2, 2022, https://ar5-syr.ipcc.ch/topic_futurechanges.php.
3. "The Time for Climate Action Is Now," BCG Executive Perspectives, April 2021, https://media-publications.bcg.com/BCG-Executive-Perspectives-Time-for-Climate-Action.pdf.
4. "Climate Finance Markets and the Real Economy," BCG & GFMA, December 2020, https://www.sifma.org/wp-content/uploads/2020/12/Climate-Finance-Markets-and-the-Real-Economy.pdf.

SECTION 1: MOMENTUM

1. In physics, momentum is "the property that a moving object has due to its mass and its motion." Merriam-Webster, accessed February 4, 2022, https://www.merriam-webster.com/dictionary/momentum.

1. PROSPERITY, WITH A CATCH

1. Robert F. Bruner and Scott Miller, *The Great Industrial Revolution in Europe: 1760–1860* (Charlottesville, VA: Darden Business Publishing, 2019).
2. "Fritz Haber, Biographical," in *Nobel Lectures, Chemistry 1901-1921* (Amsterdam: Elsevier1966), accessed February 1, 2022, https://www.nobelprize.org/prizes/chemistry/1918/haber/biographical/. First published in the book series *Les Prix Nobel*, it was later edited and republished in *Nobel Lectures*.

3. "Corn Yields in the United States, 1866 to 2014," Our World in Data, accessed February 1, 2022, https://ourworldindata.org/search?q=average+corn+yields.
4. Vaclav Smil, *Enriching the Earth: Fritz Haber, Carl Bosch, and the Transformation of World Food Production* (Cambridge, MA: MIT Press, 2000).
5. Max Roser, Hannah Ritchie, and Esteban Ortiz-Ospina, "World Population Growth," Our World in Data, last modified May 2019, https://ourworldindata.org/world-population-growth.
6. Max Roper, "Economic Growth," Our World in Data, accessed February 1, 2022, https://ourworldindata.org/economic-growth#:~:text=The%20income%20of%20the%20average%20person%20in%20the%20world%20has,times%20richer%20than%20in%201950.
7. "Real GDP Per Capita, 2017," Our World in Data, accessed February 1, 2022, https://ourworldindata.org/grapher/real-gdp-per-capita-pennwt.
8. "AR6 Climate Change 2021: The Physical Science Basis," IPCC, accessed February 1, 2022, https://www.ipcc.ch/report/ar6/wg1/#SPM.
9. "AR6 Climate Change 2021: The Physical Science Basis," IPCC.
10. Chelsea Harvey, "Earth Hasn't Warmed This Fast in Tens of Millions of Years," *E&E News, Scientific American*, September 13, 2020, https://www.scientificamerican.com/article/earth-hasnt-warmed-this-fast-in-tens-of-millions-of-years/.
11. J. Rogelj, D. Shindell, K. Jiang, et al., "Mitigation Pathways Compatible with 1.5°C in the Context of Sustainable Development," in *Global Warming of 1.5°C. An IPCC Special Report on the Impacts of Global Warming of 1.5°C Above Pre-industrial Levels and Related Global Greenhouse Gas Emission Pathways, in the Context of Strengthening the Global Response to the Threat of Climate Change, Sustainable Development, and Efforts to Eradicate Poverty*, ed. V. Masson-Delmotte, P. Zhai, H.-O. Pörtner, et al. (IPCC, 2018), https://www.ipcc.ch/site/assets/uploads/sites/2/2019/05/SR15_Chapter2_Low_Res.pdf.
12. Philip Shabecoff, "Global Warming Has Begun, Expert Tells Senate," *New York Times*, June 24, 1988, https://www.nytimes.com/1988/06/24/us/global-warming-has-begun-expert-tells-senate.html.
13. Kat Eschner, "Leaded Gas Was a Known Poison the Day It Was Invented," *Smithsonian*, December 9, 2016, https://www.smithsonianmag.com/smart-news/leaded-gas-poison-invented-180961368/.
14. Jerome O. Nriagu, "The Rise and Fall of Leaded Gasoline," *Science of the Total Environment* 92 (1990): 13–28, http://www.columbia.edu/itc/sipa/envp/louchouarn/courses/env-chem/Pb-Rise&Fall%28Nriagu1990%29.pdf.
15. "Remarks by the President on the Paris Agreement," White House, Office of the Press Secretary, October 5, 2016, https://obamawhitehouse.archives.gov/the-press-office/2016/10/05/remarks-president-paris-agreement.
16. "Emissions Gap Report 2016," UN Environment, November 26, 2016, https://www.unenvironment.org/resources/emissions-gap-report-2016.
17. "Emissions Gap Report 2020," UNEP, UNEP DTU Partnership, December 9, 2020, https://www.unep.org/emissions-gap-report-2020.
18. "Emissions Gap Report 2020," UNEP, UNEP DTU Partnership.
19. Phil Drew and Ruairidh Macintosh, "Amount of Finance Committed to Achieving 1.5°C Now at Scale Needed to Deliver the Transition," GFANZ, November 3, 2021, https://www.gfanzero.com/press/amount-of-finance-committed-to-achieving-1-5c-now-at-scale-needed-to-deliver-the-transition/.

20. Rochelle Toplensky, "Business Is the Game-Changer at COP26 in Glasgow," *Wall Street Journal*, November 6, 2021, https://www.wsj.com/articles/business-is-the-game-changer-at-cop26-in-glasgow-11636196493?st=f7gtop5mhup8a94&reflink=article_email_share.

2. INVESTING IN THE ERA OF CLIMATE CHANGE

1. Taylor Telford and Dino Grandoni, "Murray Energy Files for Bankruptcy as Coal's Role in U.S. Power Dwindles," *Washington Post*, October 29, 2019, https://www.washingtonpost.com/business/2019/10/29/coal-giant-murray-energy-files-bankruptcy-coals-role-us-power-dwindles/.
2. Pippa Stevens, "Exxon Mobil Replaced by a Software Stock After 92 Years in the Dow Is a 'Sign of the Times,'" CNBC, August 25, 2020, https://www.cnbc.com/2020/08/25/exxon-mobil-replaced-by-a-software-stock-after-92-years-in-the-dow-is-a-sign-of-the-times.html.
3. Reed Stevenson, "Tesla Overtakes Toyota as the World's Most Valuable Automaker," Bloomberg, July 1, 2020, https://www.bloomberg.com/news/articles/2020-07-01/tesla-overtakes-toyota-as-the-world-s-most-valuable-automaker?sref=q3MO9qbb.
4. Mike Murphy, "Beyond Meat Soars 163 Percent in Biggest-Popping U.S. IPO Since 2000," MarketWatch, May 5, 2019, https://www.marketwatch.com/story/beyond-meat-soars-163-in-biggest-popping-us-ipo-since-2000-2019-05-02.
5. Cary Funk and Brian Kennedy, "How Americans See Climate Change and the Environment in 7 Charts," Pew Research Center, April 21, 2020, https://www.pewresearch.org/fact-tank/2020/04/21/how-americans-see-climate-change-and-the-environment-in-7-charts/.
6. Matthew Ballew, Jennifer Marlon, John Kotcher, et al., "Climate Note: Young Adults, Across Party Lines, Are More Willing to Take Climate Action," Yale Program on Climate Change Communication, April 28, 2020, https://climatecommunication.yale.edu/publications/young-adults-climate-activism/.
7. Audrey Choi, "How Younger Investors Could Reshape the World," Morgan Stanley Wealth Management, January 24, 2018, https://www.morganstanley.com/access/why-millennial-investors-are-different.
8. "Amazon Sustainability," Amazon.com, accessed February 4, 2022, https://sustainability.aboutamazon.com/.
9. Tom Murray, "Apple, Ford, McDonald's and Microsoft Among This Summer's Climate Leaders," Environmental Defense Fund, August 10, 2020, https://www.edf.org/blog/2020/08/10/apple-ford-mcdonalds-and-microsoft-among-summers-climate-leaders.
10. "Microsoft Announces It Will Be Carbon Negative by 2030," Microsoft News Center, January 16, 2020, https://news.microsoft.com/2020/01/16/microsoft-announces-it-will-be-carbon-negative-by-2030/#:~:text=%E2%80%9CBy%202030%20Microsoft%20will%20be,it%20was%20founded%20in%201975.%E2%80%9D.
11. Heleen L. van Soest, Michel G. J. den Elzen, and Detlef P. van Vuuren, "Net-Zero Emission Targets for Major Emitting Countries Consistent with the Paris Agreement," *Nature Communications* 12 (April 9, 2021): article 2140, https://www.nature.com/articles/s41467-021-22294-x.
12. *The Long-Term Strategy of the United States: Pathways to Net-Zero Greenhouse Gas Emissions by 2050* (Washington, DC: U.S. Department of State and U.S. Executive Office

of the President, 2021), https://www.whitehouse.gov/wp-content/uploads/2021/10/US-Long-Term-Strategy.pdf.
13. "State Renewable Portfolio Standards and Goals," NCSL, August 13, 2021, https://www.ncsl.org/research/energy/renewable-portfolio-standards.aspx#:~:text=489%20.
14. "How did you go bankrupt?" "Two ways. Gradually and then suddenly," Quote Investigator, accessed February 4, 2022, https://quoteinvestigator.com/2018/08/06/bankrupt/.

3. MOMENTUM

1. Larry Fink's 2020 Letter to CEOs, BlackRock, https://www.blackrock.com/corporate/investor-relations/2020-larry-fink-ceo-letter.
2. David Solomon, "Goldman Sachs' Commercially Driven Plan for Sustainability," *Financial Times*, December 15, 2019, https://www.ft.com/content/ffd794c8-183a-11ea-b869-0971bffac109.
3. Marcie Frost, "How California's Pension Fund Is Confronting Climate Change," CalMatters, last modified December 17, 2019, https://calmatters.org/commentary/2019/12/pension-climate-change/.
4. Leslie Hook and Gillian Tett, "Hedge Fund TCI Vows to Punish Directors Over Climate Change," *Financial Times*, December 1, 2019, https://www.ft.com/content/dde5e4d4-140f-11ea-9ee4-11f260415385.
5. "We need to reach net zero emissions by 2050," Financing Roadmaps, accessed February 4, 2022, https://www.gfanzero.com/netzerofinancing.

SECTION 2: CLIMATE CHANGE SOLUTIONS

1. *Special Report: Global Warming of 1.5 °C*, IPCC, accessed February 4, 2022, https://www.ipcc.ch/sr15/.
2. Michelle Della Vigna, Zoe Stavrinou, and Alberto Gandolfi, "Carbonomics: Innovation, Deflation and Affordable De-carbonization," Goldman Sachs, October 13, 2020, https://www.goldmansachs.com/insights/pages/gs-research/carbonomics-innovation-deflation-and-affordable-de-carbonization/report.pdf.
3. "GDP (Current US$)," World Bank, accessed February 4, 2022, https://data.worldbank.org/indicator/NY.GDP.MKTP.CD.

4. RENEWABLE ENERGY

The Edison quote on the chapter-opening page is taken from Heather Rogers, "Reconsideration: Current Thinking," *New York Times*, June 3, 2007, https://www.nytimes.com/2007/06/03/magazine/03wwln-essay-t.html#:~:text=In%201931%2C%20not%20long%20before,out%20before%20we%20tackle%20that.%E2%80%9D.
1. Material in this section is sourced from Bruce Usher, *Renewable Energy: A Primer for the Twenty-First Century* (New York: Columbia University Press, 2019), chap. 5.
2. "How Does Solar Work?" Office of Energy Efficiency & Renewable Energy, accessed February 4, 2022, https://www.energy.gov/eere/solar/how-does-solar-work.

3. "The Nobel Prize in Physics 1921, Albert Einstein," Nobelprize.org, accessed February 4, 2022, https://www.nobelprize.org/nobel_prizes/physics/laureates/1921/.
4. Jeremy Hsu, "Vanguard 1, First Solar-Powered Satellite, Still Flying at 50," Space .com, March 18, 2008, https://www.space.com/5137-solar-powered-satellite-flying-50 .html.
5. "Photovoltaic Energy Factsheet," Center for Sustainable Systems, University of Michigan, 2017, http://css.umich.edu/factsheets/photovoltaic-energy-factsheet.
6. Ramez Naam, "Smaller, Cheaper, Faster: Does Moore's Law Apply to Solar Cells?" *Scientific American*, March 16, 2011, https://blogs.scientificamerican.com/guest-blog /smaller-cheaper-faster-does-moores-law-apply-to-solar-cells/.
7. Martin Schachinger, "Module Price Index," *PV Magazine*, last modified January 15, 2021, https://www.pv-magazine.com/module-price-index/.
8. A different solar technology, concentrated solar power (CSP), uses mirrors to concentrate the sun's rays. The heat is used to drive a steam-powered turbine to generate electricity. However, concentrated solar power has fallen out of favor because the cost of generating electricity from CSP is significantly higher than from PV, making it less competitive.
9. Annie Sneed, "Moore's Law Keeps Going, Defying Expectations," *Scientific American*, May 19, 2015, https://www.scientificamerican.com/article/moore-s-law-keeps-going -defying-expectations/.
10. "Sunny Uplands," Science & Technology, *The Economist*, November 21, 2012, https:// www.economist.com/news/2012/11/21/sunny-uplands.
11. "International Technology Roadmap for Photovoltaic (ITRPV) 2020 Results," 12th ed., April 2021, https://www.vdma.org/international-technology-roadmap-photovoltaic.
12. "International Technology Roadmap for Photovoltaic (ITRPV) 2020 Results," 12 ed.
13. "Lazard's Levelized Cost of Energy Analysis—Version 14.0," Lazard, October 2020, https://www.lazard.com/media/451419/lazards-levelized-cost-of-energy-version-140 .pdf.
14. Billy Ludt, "What Is a Solar Tracker and How Does It Work?" Solar Power World, January 16, 2020, https://www.solarpowerworldonline.com/2020/01/what-is-a-solar -tracker-and-how-does-it-work/.
15. "Bifacial Solar Advances with the Times—and the Sun," NREL, accessed February 4, 2022, https://www.nrel.gov/news/features/2020/bifacial-solar-advances-with-the-times -and-the-sun.html.
16. "Solar Industry Research Data," SEIA, accessed February 4, 2022, http://www.seia.org /research-resources/solar-industry-data.
17. "WoodMac Expects Up to 25 Percent Drop in Solar Costs This Decade," Renewables Now, January 21, 2021, https://renewablesnow.com/news/woodmac-expects-up-to-25 -drop-in-solar-costs-this-decade-728679/.
18. Andrew Z. P. Smith, "Fact Checking Elon Musk's Blue Square: How Much Solar to Power the US?" UCL Energy Institute blog, May 21, 2015, https://blogs.ucl.ac.uk /energy/2015/05/21/fact-checking-elon-musks-blue-square-how-much-solar-to -power-the-us/.
19. "News Release: NREL Raises Rooftop Photovoltaic Technical Potential Estimate," NREL, March 24, 2016, https://www.nrel.gov/news/press/2016/24662.html.
20. *The Energy Outlook*, 2020 ed., BP, https://www.bp.com/content/dam/bp/business -sites/en/global/corporate/pdfs/energy-economics/energy-outlook/bp-energy -outlook-2020.pdf.

21. "Amazon Becomes World's Largest Corporate Purchaser of Renewable Energy, Advancing Its Climate Pledge Commitment to Be Net-Zero Carbon by 2040," BusinessWire, December 10, 2020, https://www.businesswire.com/news/home/20201210005304/en/Amazon-Becomes-World%E2%80%99s-Largest-Corporate-Purchaser-of-Renewable-Energy-Advancing-its-Climate-Pledge-Commitment-to-be-Net-zero-Carbon-by-2040#:~:text=%E2%80%9CWith%20a%20total%20of%20127,our%20original%20target%20of%202030.
22. Emma Foehringer Merchant, "California's Rooftop Solar Mandate Hits Snag with Housing Market Set for Slowdown," GTM: A Wood Mackenzie Business, July 22, 2020, https://www.greentechmedia.com/articles/read/will-the-coronavirus-slow-californias-solar-home-requirement.
23. Jon Moore and Seb Henbest, "New Energy Outlook 2020," BloombergNEF, October 2020, https://assets.bbhub.io/professional/sites/24/928908_NEO2020-Executive-Summary.pdf.
24. Material in this section is sourced from Usher, *Renewable Energy*, chap. 4.
25. "An Industry First: Haliade-X Offshore Wind Turbine," GE Renewable Energy, accessed February 4, 2022, https://www.ge.com/renewableenergy/wind-energy/offshore-wind/haliade-x-offshore-turbine.
26. "Lazard's Levelized Cost of Energy Analysis—Version 15.0," Lazard, October 2021, https://www.lazard.com/media/451905/lazards-levelized-cost-of-energy-version-150-vf.pdf.
27. "Wind Explained: Electricity Generation from Wind," EIA, last modified March 17, 2021, https://www.eia.gov/energyexplained/wind/electricity-generation-from-wind.php.
28. "Wind Energy Basics," NYSERDA, accessed February 4, 2022, https://www.nyserda.ny.gov/-/media/Files/Publications/Research/Biomass-Solar-Wind/NY-Wind-Energy-Guide-1.pdf.
29. Jan Dell and Matthew Klippenstein, "Wind Power Could Blow Past Hydro's Capacity Factor by 2020," GTM: A Wood Mackenzie Business, February 8, 2017, https://www.greentechmedia.com/articles/read/wind-power-could-blow-past-hydros-capacity-factor-by-2020.
30. Jason Finkelstein, David Frankel, and Jesse Noffsinger, "How to Decarbonize Global Power Systems," McKinsey & Company, May 19, 2020, https://www.mckinsey.com/industries/electric-power-and-natural-gas/our-insights/how-to-decarbonize-global-power-systems.
31. Graeme R. G. Hoste, Michael J. Dvorak, and Mark Z. Jacobson, "Matching Hourly and Peak Demand by Combining Different Renewable Energy Sources: A Case Study for California in 2020," Stanford University, Department of Civil and Environmental Engineering, accessed February 4, 2022, https://web.stanford.edu/group/efmh/jacobson/Articles/I/CombiningRenew/HosteFinalDraft.
32. Justin Gerdes, "California's Wind Market Has All But Died Out. Could Grid Services Revenue Help?" GTM: A Wood Mackenzie Business, March 30, 2020, https://www.greentechmedia.com/articles/read/justin-california.
33. "Wind Turbines," U.S. Fish & Wildlife Services, last modified April 18, 2018, https://www.fws.gov/birds/bird-enthusiasts/threats-to-birds/collisions/wind-turbines.php.
34. "Cats Indoors," American Bird Conservancy, accessed February 4, 2022, https://abcbirds.org/program/cats-indoors/cats-and-birds/.
35. National Audubon Society, "Climate: Wind Power and Birds," July 21, 2020, https://www.audubon.org/news/wind-power-and-birds.

36. Jason Samenow, "Blowing Hard: The Windiest Time of Year," *Washington Post*, March 31, 2016, https://www.washingtonpost.com/news/capital-weather-gang/wp/2014/03/26/what-are-the-windiest-states-and-cities-what-is-d-c-s-windiest-month/?utm_term=.4ac52623e129.
37. "Offshore Wind Outlook 2019," IEA, November 2019, https://www.iea.org/reports/offshore-wind-outlook-2019.
38. "An Industry First," GE Renewable Energy.
39. "Block Island Wind Farm," Power Technology, December 30, 2016, https://www.power-technology.com/projects/block-island-wind-farm/.
40. "Offshore Wind Projects," New York State ERDA, November 19, 2021, https://www.nyserda.ny.gov/All-Programs/Programs/Offshore-Wind/Focus-Areas/NY-Offshore-Wind-Projects.
41. "Levelized Cost of Energy, Levelized Cost of Storage, and Levelized Cost of Hydrogen 2020," Lazard, October 19, 2020, https://www.lazard.com/perspective/levelized-cost-of-energy-and-levelized-cost-of-storage-2020/.
42. "Governor Cuomo Announces Finalized Contracts for Empire Wind and Sunrise Wind Offshore Wind Projects to Deliver Nearly 1,700 Megawatts of Clean and Affordable Renewable Energy to New Yorkers," New York State ERDA, November 19, 2021, https://www.nyserda.ny.gov/About/Newsroom/2019-Announcements/2019-10-23-Governor-Cuomo-Announces-Finalized-Contracts-for-Empire-Wind-and-Sunrise-Wind-Offshore-Wind-Projects.
43. Eric Paya and Aaron Zigeng Du, "The Frontier Between Fixed and Floating Foundations in Offshore Wind," Empire Engineering, October 19, 2020, https://www.empireengineering.co.uk/the-frontier-between-fixed-and-floating-foundations-in-offshore-wind/.
44. Sarah McFarlane, "Floating Wind Turbines Buoy Hopes of Expanding Renewable Energy," The Future of Everything: Energy & Climate, *Wall Street Journal*, February 6, 2021, https://www.wsj.com/articles/floating-wind-turbines-buoy-hopes-of-expanding-renewable-energy-11612623702?mod=article_inline.
45. McFarlane, "Floating Wind Turbines."
46. McFarlane, "Floating Wind Turbines."
47. Moore and Henbest, "New Energy Outlook 2020."
48. Gregory Meyer, "Offshore Wind Power Project Moves Ahead Under Biden," *Financial Times*, March 8, 2021, https://www.ft.com/content/923de0ae-b72e-4106-9688-c549b6bb1690.
49. "FAQ: How Old Are U.S. Nuclear Power Plants, and When Was the Newest One Built?" EIA, last modified December 29, 2020, https://www.eia.gov/tools/faqs/faq.php?id=228&t=21#:~:text=The%20newest%20reactor%20to%20enter,nuclear%20reactors%20for%2040%20years.
50. Julian Spector, "Sole US Nuclear Plant Under Construction Plods on Despite Virus Infections," GTM: A Wood Mackenzie Business, April 30, 2020, https://www.greentechmedia.com/articles/read/covid-19-impacted-productivity-of-vogtle-nuclear-plant-construction.
51. "Lazard's Levelized Cost of Energy Analysis—Version 14.0," Lazard.
52. "Annual Energy Outlook 2020, with Projections to 2050," EIA, January 29, 2020, https://www.eia.gov/outlooks/aeo/pdf/AEO2020%20Full%20Report.pdf.
53. Adrian Cho, "U.S. Department of Energy Rushes to Build Advanced Nuclear Reactors," News, *Science*, May 20, 2020, https.www.sciencemag.org/news/2020/05/us-department-energy-rushes-build-advanced-new-nuclear-reactors.

54. Catherine Clifford, "Bill Gates: Nuclear Power Will 'Absolutely' Be Politically Acceptable Again—It's Safer than Oil, Coal, Natural Gas," CNBC, February 25, 2021, https://www.cnbc.com/2021/02/25/bill-gates-nuclear-power-will-absolutely-be-politically-acceptable.html.
55. "A Cost-Competitive Nuclear Power Solution," Nuscale, 2021, https://www.nuscalepower.com/benefits/cost-competitive#:~:text=The%20first%20module%20will%20be,new%20frontier%20of%20clean%20energy.
56. "Levelized Costs of New Generation Resources in the *Annual Energy Outlook 2021*," EIA, February 2021, https://www.eia.gov/outlooks/aeo/pdf/electricity_generation.pdf.
57. Daniel Michaels, "Mini Nuclear Reactors Offer Promise of Cheaper, Clean Power," The Future of Everything: Energy & Climate, *Wall Street Journal*, last modified February 11, 2021, https://www.wsj.com/articles/mini-nuclear-reactors-offer-promise-of-cheaper-clean-power-11613055608?&mod=article_inline.
58. Stephen Lacey, "Investors Funnel $1.3 Billion Into the Advanced Nuclear Industry," GTM: A Wood Mackenzie Business, June 17, 2015, https://www.greentechmedia.com/articles/read/investors-pour-1-3-billion-into-the-advanced-nuclear-industry.
59. "Advanced Nuclear Energy Projects Loan Guarantees," U.S. Department of Energy, Loan Programs Office, accessed February 4, 2022, https://www.energy.gov/lpo/advanced-nuclear-energy-projects-loan-guarantees.
60. "Bill Gates: Chairman of the Board," TerraPower, accessed February 4, 2022, https://www.terrapower.com/people/bill-gates/.
61. "Electricity Explained: Electricity in the United States," EIA, last modified March 18, 2021, https://www.eia.gov/energyexplained/electricity/electricity-in-the-us.php.
62. Kelly Pickerel, "Utility-Scale Solar Makes Up Nearly 30% of New U.S. Electricity Generation in 2020," Solar Power World, February 9, 2021, https://www.solarpowerworldonline.com/2021/02/utility-scale-solar-makes-up-nearly-30-of-new-u-s-electricity-generation-in-2020/#:~:text=Webinars%20%2F%20Digital%20Events-,Utility%2Dscale%20solar%20makes%20up%20nearly%2030%25%20of%20new%20U.S.,78%25%20of%202020%20electricity%20additions.
63. Julia Gheorghiu and Dive Brieff: "El Paso Electric Sees Record Low Solar Prices as It Secures New Mexico Project Approvals," Utility Dive, May 18, 2020, https://www.utilitydive.com/news/el-paso-electric-sees-record-low-solar-prices-as-it-secures-new-mexico-proj/578113/.
64. "Lazard's Levelized Cost of Energy Analysis—Version 14.0," Lazard.
65. Geoffrey Heal, "Economic Aspects of the Energy Transition," NBER, September 2020, https://www.nber.org/papers/w27766.
66. Katherine Walla, "The World Is About to Embark on a Big Energy Transition. Here's What It Could Look Like," Atlantic Council, January 19, 2021, https://www.atlanticcouncil.org/blogs/new-atlanticist/the-world-is-about-to-embark-on-a-big-energy-transition-heres-what-it-could-look-like/.
67. Bill Bostock, "The UK Has Gone 2 Months Without Burning Coal, the longest Period Since the Dawn of the Industrial Revolution," Business Insider, June 13, 2020, https://www.businessinsider.com/britain-no-coal-burning-first-time-since-industrial-revolution-2020-6.
68. U.S. Energy Information Administration, *Electric Power Annual*, October 2021, https://www.eia.gov/electricity/annual/pdf/epa.pdf.
69. Fred Pearce, "As Investors and Insurers Turn Away, the Economics of Coal Turn Toxic," Yale Environment 360, March 10, 2020, https://e360.yale.edu/features/as-investors-and-insurers-back-away-the-economics-of-coal-turn-toxic.

70. Valerie Volcovici, "Murray Energy Files for Bankruptcy as U.S. Coal Decline Continues," Reuters, October 29, 2019, https://www.reuters.com/article/us-usa-coal-bankruptcy/murray-energy-files-for-bankruptcy-as-u-s-coal-decline-continues-idUSKBN1X81SB.
71. Climate Solutions/Energy Solutions: "Natural Gas," C2ES, Center for Climate and Energy Solutions, accessed February 4, 2022, https://www.c2es.org/content/natural-gas/.
72. "Lazard's Levelized Cost of Energy Analysis—Version 14.0," Lazard.
73. "Electricity Explained," EIA.
74. Dennis Wamsted, "IEEFA U.S.: Utilities Are Now Skipping the Gas 'Bridge' in Transition from Coal to Renewables," Institute for Energy Economics and Financial Analysis (IEEFA), July 1, 2020, https://ieefa.org/ieefa-u-s-utilities-are-now-skipping-the-gas-bridge-in-transition-from-coal-to-renewables/.
75. Dennis Wamsted, "IEEFA U.S."
76. "Electricity Explained," EIA.
77. "Annual Energy Outlook 2021, with Projections to 2050," EIA, February 2021, https://www.eia.gov/outlooks/aeo/pdf/AEO_Narrative_2021.pdf.
78. IEA, *Geothermal Power* (Paris: IEA, 2021), https://www.iea.org/reports/geothermal.
79. "Global Energy Perspective 2021," McKinsey & Company, January 2021, https://www.mckinsey.com/~/media/McKinsey/Industries/Oil%20and%20Gas/Our%20Insights/Global%20Energy%20Perspective%202021/Global-Energy-Perspective-2021-final.pdf.
80. *The Energy Outlook*, 2020 ed., BP.
81. Michelle Della Vigna, Zoe Stavrinou, and Alberto Gandolfi, "Carbonomics: The Green Engine of Economic Recovery," Goldman Sachs, Equity Research, June 16, 2020, https://www.goldmansachs.com/insights/pages/gs-research/carbonomics-green-engine-of-economic-recovery-f/report.pdf.

5. ELECTRIC VEHICLES

Material in this chapter is sourced from Bruce Usher, *Renewable Energy: A Primer for the Twenty-First Century* (New York: Columbia University Press, 2019), chap. 8.

1. Robert L. Bradley Jr., "Electric Vehicles: As in 1896, the Wrong Way to Go," IER, October 19, 2010, http://instituteforenergyresearch.org/analysis/electric-vehicles-as-in-1896-the-wrong-way-to-go/.
2. RLF Attorneys, "Who Got America's First Speeding Ticket?" Rosenblum Law, June 20, 2016, http://newyorkspeedingfines.com/americas-speeding-ticket/.
3. Derek Markham, "This $10K Air-Powered Vehicle Could Be the Tiny Car to Go with Your Tiny House," Treehugger, last modified October 11, 2018, https://www.mnn.com/green-tech/transportation/blogs/porsches-long-buried-first-vehicle-was-an-electric-car-and-it-was.
4. "All Electric Vehicles," U.S. Department of Energy, accessed February 5, 2022, https://www.fueleconomy.gov/feg/evtech.shtml.
5. Dan Strohl, "Ford, Edison and the Cheap EV That Almost Was," *Wired*, June 18, 2010, https://www.wired.com/2010/06/henry-ford-thomas-edison-ev/.
6. Martin V. Melosi, "The Automobile and the Environment in American History," Automobile in American Life and Society, accessed February 5, 2022, http://www.autolife.umd.umich.edu/Environment/E_Overview/E_Overview3.htm.
7. Bob Casey, "Past Forward, Activating the Henry Ford Archive of Innovation," The Henry Ford, June 22, 2015, https://www.thehenryford.org/explore/blog/general-motors-ev1/.

8. Elon Musk, "The Secret Tesla Motors Master Plan (just between you and me)," Tesla, August 2, 2006, https://www.tesla.com/blog/secret-tesla-motors-master-plan-just-between-you-and-me.
9. Electric vehicles receive a miles-per-gallon *equivalent* rating for consumers to compare fuel efficiency against gasoline-powered vehicles.
10. Kim Reynolds, "2008 Tesla Roadster First Drive," *MotorTrend*, January 23, 2008, http://www.motortrend.com/cars/tesla/roadster/2008/2008-tesla-roadster/.
11. Bjorn Nykvist and Mans Nilsson, "Rapidly Falling Costs of Battery Packs for Electric Vehicles," *Nature Climate Change*, February 9, 2015, https://mediamanager.sei.org/documents/Publications/SEI-Nature-pre-pub-2015-falling-costs-battery-packs-BEVs.pdf.
12. "Federal Tax Credits for New All-Electric and Plug-in Hybrid Vehicles," U.S. Department of Energy, last modified November 3, 2021, https://www.fueleconomy.gov/feg/taxevb.shtml.
13. Rob Wile, "Credit Suisse Gives Point-by-Point Breakdown Why Tesla Is Better than Your Regular Car," Business Insider, August 14, 2014, http://www.businessinsider.com/credit-suisse-on-tesla-2014-8.
14. Leslie Shaffer, "JPMorgan Thinks the Electric Vehicle Revolution Will Create a Lot of Losers," CNBC, August 22, 2017, https://www.cnbc.com/2017/08/22/jpmorgan-thinks-the-electric-vehicle-revolution-will-create-a-lot-of-losers.html.
15. Steven Szakaly and Patrick Manzi, "Nada Data 2015," Nada, accessed February 5, 2022, https://www.nada.org/WorkArea/DownloadAsset.aspx?id=21474839497.
16. Dana Hull, "Tesla Said It Received Over 325,000 Model 3 Reservations," Bloomberg, April 7, 2016, https://www.bloomberg.com/news/articles/2016-04-07/tesla-says-model-3-pre-orders-surge-to-325-000-in-first-week?sref=3rbSWFkc.
17. "Global Top 20—December 2020," EV Sales, February 2, 2021, http://ev-sales.blogspot.com/2021/02/global-top-20-december-2020.html.
18. Peter Campbell, "Electric Car Rivals Revved Up to Challenge Tesla," *Financial Times*, September 21, 2018, https://www.ft.com/content/3f5ded00-bd7d-11e8-8274-55b72926558f.
19. Srikant Inampudi, Nicolaas Kramer, Inga Maurer, and Virginia Simmons, "As Dramatic Disruption Comes to Automotive Showrooms, Proactive Dealers Can Benefit Greatly," McKinsey & Company, January 23, 2019, https://www.mckinsey.com/industries/automotive-and-assembly/our-insights/as-dramatic-disruption-comes-to-automotive-showrooms-proactive-dealers-can-benefit-greatly.
20. Steven Loveday, "US EV Sales Hit All-Time High in Q4 2021: Tesla Leads w/72 % Share," InsideEVs, February 24, 2022, https://insideevs.com/news/569711/tesla-leads-ev-sales-surge/.
21. 24/7 Wall St., "How Many Gas Stations Are in U.S.? How Many Will There Be in 10 Years?" MarketWatch, February 16, 2020, https://www.marketwatch.com/story/how-many-gas-stations-are-in-us-how-many-will-there-be-in-10-years-2020-02-16.
22. Tina Bellon and Paul Lienert, "Change Suite: Factbox: Five Facts on the State of the U.S. Electric Vehicle Charging Network," Reuters, September 1, 2021, https://www.reuters.com/world/us/five-facts-state-us-electric-vehicle-charging-network-2021-09-01/#:~:text=The%20United%20States%20currently%20has,U.S.%20Department%20of%20Energy%20data.
23. Rebecca Lindland, "How Long Does It Take to Charge an Electric Car?" J. D. Power, March 26, 2020, https://www.jdpower.com/cars/shopping-guides/how-long-does-it-take-to-charge-an-electric-car.

24. "Battery Pack Prices Cited Below $100/kWh for the First Time in 2020, While Market Average Sits at $137/kWh," BloombergNEF, December 16, 2020, https://about.bnef.com/blog/battery-pack-prices-cited-below-100-kwh-for-the-first-time-in-2020-while-market-average-sits-at-137-kwh/.
25. Akshat Rathii, "The Magic Number That Unlocks the Electric Car Revolution," Bloomberg News, September 22, 2020, https://www.bloomberg.com/news/articles/2020-09-22/elon-musk-s-battery-day-could-reveal-very-cheap-batteries.
26. Jack Ewing and Ivan Penn, "The Auto Industry Bets Its Future on Batteries," *New York Times*, last modified May 4, 2021, https://www.nytimes.com/2021/02/16/business/energy-environment/electric-car-batteries-investment.html?action=click&module=Well&pgtype=Homepage§ion=Business.
27. Aarian Marshall, "The Intersection Between Self-Driving Cars and Electric Cars," *Wired*, July 13, 2020, https://www.wired.com/story/intersection-self-driving-cars-electric/.
28. Neal E. Boudette and Coral Davenport, "G.M. Will Sell Only Zero-Emission Vehicles by 2035," *New York Times*, last modified October 1, 2021, https://www.nytimes.com/2021/01/28/business/gm-zero-emission-vehicles.html.
29. Akshat Rathi, "If Tesla Is the Apple of Electric Vehicles, Volkswagen Is Betting It Can Be Samsung," Bloomberg Green, March 26, 2021, https://www.bloomberg.com/news/articles/2021-03-16/if-tesla-is-the-apple-of-electric-vehicles-volkswagen-is-betting-it-can-be-samsung?sref=3rbSWFkc.
30. Jack Denton, "Forget Nio and XPeng. This company and Tesla will be the top two electric-vehicle plays by 2025, says UBS," MarketWatch, March 13, 2021, https://www.marketwatch.com/story/forget-nio-and-xpeng-this-company-and-tesla-will-be-the-top-2-electric-vehicle-plays-by-2025-says-ubs-11615306959.
31. "Sources of Greenhouse Gas Emissions," EPA, accessed February 5, 2022, https://www.epa.gov/ghgemissions/sources-greenhouse-gas-emissions#transportation.

6. ENERGY STORAGE

The quote in the chapter-opening image caption is taken from Brian Eckhouse and David Stringer, "A Megabattery Boom Is Coming to Rescue Overloaded Power Grids," Bloomberg Businessweek, January 22, 2021, https://www.bloomberg.com/news/articles/2021-01-22/megabattery-boom-will-rescue-overloaded-power-grids?sref=3rbSWFkc.
1. Material in this chapter is sourced from Bruce Usher, *Renewable Energy: A Primer for the Twenty-First Century* (New York: Columbia University Press, 2019), chap. 10.
2. "What the Duck Curve Tells Us About Managing a Green Grid," California ISO, 2016, https://www.caiso.com/Documents/FlexibleResourcesHelpRenewables_FastFacts.pdf.
3. "Phase Out Peakers," Clean Energy Group, accessed February 5, 2022, https://www.cleanegroup.org/ceg-projects/phase-out-peakers/.
4. "Lazard's Levelized Cost of Energy Analysis—Version 14.0," Lazard, October 2020, https://www.lazard.com/media/451419/lazards-levelized-cost-of-energy-version-140.pdf.
5. Andrew Blakers Matthew Stocks, Bin Lu, and Cheng Cheng, "A Review of Pumped Hydro Energy Storage," *Progress in Energy* 3, no. 2 (March 25, 2021): 022003, https://iopscience.iop.org/article/10.1088/2516-1083/abeb5b.
6. "Packing Some Power," *The Economist*, March 3, 2012, http://www.economist.com/node/21548495?frsc=dg%7Ca.

6. ENERGY STORAGE

7. "Lazard's Levelized Cost of Storage—Version 2.0," Lazard, December 2016, https://www.lazard.com/media/438042/lazard-levelized-cost-of-storage-v20.pdf.
8. Thomas Fisher, "Tesla Alone Could Double Global Demand for the Laptop Batteries It Uses," Reuters, September 4, 2013, https://www.reuters.com/article/idUS302095204320130905.
9. Adele Peters, "Inside Tesla's 100% Renewable Design for the Gigafactory," Fast Company, April 15, 2019, https://www.fastcompany.com/90334858/inside-teslas-100-renewable-design-for-the-Gigafactory.
10. Bruce Usher and Geoff Heal, "Architects of the Future? Tesla, Inc., Energy, Transportation, and the Climate," Columbia CaseWorks, August 11, 2020.
11. Micah S. Ziegler and Jessika E. Trancik, "Re-examining Rates of Lithium-Ion Battery Technology Improvement and Cost Decline," *Energy and Environmental Science* 4 (March 23, 2021), https://pubs.rsc.org/en/content/articlelanding/2021/EE/D0EE02681F#!divAbstract.
12. Phil LeBeau, "Tesla's Lead in Batteries Will Last Through Decade While GM Closes In," CNBC, March 10, 2021, https://www.cnbc.com/2021/03/10/teslas-lead-in-batteries-will-last-through-decade-while-gm-closes-in-.html.
13. Julian Spector, "Tesla Battery Day: Expect Battery Costs to Drop by Half Within 3 Years," GTM: A Wood Mackenzie Business, September 22, 2020, https://www.greentechmedia.com/articles/read/tesla-battery-day-cost-reduction-three-years.
14. Battery energy storage systems are characterized by rated power (MW) and energy storage capacity (MWh). For example, a 50 MW system for 4 hours has 200 MWh of electrical output.
15. "Vistra Brings World's Largest Utility-Scale Battery Energy Storage System Online," Cision: PR Newswire, January 26, 2021, https://www.prnewswire.com/news-releases/vistra-brings-worlds-largest-utility-scale-battery-energy-storage-system-online-301202027.html.
16. "Energy Storage Grand Challenge: Energy Storage Market Report," U.S. Department of Energy, December 2020, https://www.energy.gov/sites/prod/files/2020/12/f81/Energy%20Storage%20Market%20Report%202020_0.pdf.
17. Zhao Liu, "The History of the Lithium-Ion Battery," ThermoFisher Scientific, October 11, 2019, https://www.thermofisher.com/blog/microscopy/the-history-of-the-lithium-ion-battery/.
18. "The Truck of the Future Is Here: All-Electric Ford F-150," Ford, May 19, 2021, https://media.ford.com/content/fordmedia/fna/us/en/news/2021/05/19/all-electric-ford-f-150-lightning.html.
19. Isobel Asher Hamilton, "Tesla Is Letting California Solar Customers with Powerwalls Feed Their Energy Back into the Grid to Help Prevent Blackouts," Business Insider, July 23, 2021, https://www.businessinsider.com/tesla-powerwall-virtual-power-plant-california-grid-solar-energy-2021-7.
20. "2035 The Report," Goldman School of Public Policy, University of California Berkeley, June 2020, http://www.2035report.com/wp-content/uploads/2020/06/2035-Report.pdf?hsCtaTracking=8a85e9ea-4ed3-4ec0-b4c6-906934306ddb%7Cc68c2ac2-1db0-4d1c-82a1-65ef4daaf6c1.
21. "Annual Energy Outlook 2020," EIA, accessed February 5, 2022, https://www.eia.gov/outlooks/aeo/data/browser/#/?id=9-AEO2020&cases=ref2020&sourcekey=0.

7. GREEN HYDROGEN

The quote on the chapter-opening page is taken from Christopher M. Matthews and Katherine Blunt, "Green Hydrogen Plant in Saudi Desert Aims to Amp Up Clean Power," *Wall Street Journal*, February 8, 2021, https://www.wsj.com/articles/green-hydrogen-plant-in-saudi-desert-aims-to-amp-up-clean-power-11612807226.

1. "Path to Hydrogen: A Cost Perspective," Hydrogen Council, January 20, 2020, https://hydrogencouncil.com/wp-content/uploads/2020/01/Path-to-Hydrogen-Competitiveness_Full-Study-1.pdf.
2. "Green Hydrogen: The Next Transformational Driver of the Utilities Industry," Goldman Sachs, September 22, 2020, https://www.goldmansachs.com/insights/pages/gs-research/green-hydrogen/report.pdf.
3. Highlights compiled by Kimmie Skinner and Celine Yang, "Congressional Climate Camp #2: Federal Policies for High Emitting Sectors, EESI," EESI, February 26, 2021, https://www.eesi.org/briefings/view/022621camp.
4. "Airbus Reveals New Zero-Emission Concept Aircraft," Airbus, September 21, 2020, https://www.airbus.com/newsroom/press-releases/en/2020/09/airbus-reveals-new-zeroemission-concept-aircraft.html.
5. "Transitioning to Green Fertilizers in Agriculture: Outlook and Opportunities," University of Minnesota, ARPA-E Macroalgae Valorization Workshop, November 16, 2020, https://arpa-e.energy.gov/sites/default/files/Mon5%20Reese%20-%20Transition%20to%20Green%20Fertilizer%20-%20ARPA-E%20Macroalgae%20Workshop%20FINAL.pdf.
6. Anmar Frangoul, "Sweden Will Soon Be Home to a Major Steel Factory Powered by the 'World's Largest Green Hydrogen Plant,'" *Sustainable Energy*, CNBC, February 25, 2021, https://www.cnbc.com/2021/02/25/steel-factory-to-be-powered-by-worlds-largest-green-hydrogen-plant.html.
7. Gerson Freitas Jr. and Chris Martin, "Cheap Wind Power Could Boost Green Hydrogen," Bloomberg Green, July 24, 2020, https://origin.www.bloomberg.com/news/articles/2020-07-23/cheap-wind-power-could-boost-green-hydrogen-morgan-stanley-says?cmpid=BBD072420_GREENDAILY&utm_medium=email&utm_source=newsletter&utm_term=200724&utm_campaign=greendaily&sref=q3MO9qbb.
8. "Hydrogen Economy Outlook: Key Messages," BloombergNEF, March 30, 2020, https://data.bloomberglp.com/professional/sites/24/BNEF-Hydrogen-Economy-Outlook-Key-Messages-30-Mar-2020.pdf.
9. John Parnell, "World's Largest Green Hydrogen Project Unveiled in Saudi Arabia," GTM: A Wood Mackenzie Business, July 7, 2020, https://www.greentechmedia.com/articles/read/us-firm-unveils-worlds-largest-green-hydrogen-project.
10. "Hydrogen: Beyond the Hype," S&P Global: Platts, accessed February 5, 2022, https://www.spglobal.com/platts/en/market-insights/topics/hydrogen.
11. Parnell, "World's Largest Green Hydrogen Project."
12. Andrew Moore, "Air Products Expects 'First-Mover' Advantage with Ambitious Saudi Arabia Hydrogen Project," S&P Global: Platts, September 30, 2020, https://www.spglobal.com/platts/en/market-insights/latest-news/electric-power/093020-air-products-expects-first-mover-advantage-with-ambitious-saudi-arabia-hydrogen-project.

13. "Hydrogen Economy Outlook: Key Messages," BloombergNEF.
14. Blake Matich, "Global Hydrogen Project Pipeline Expected to Exceed $300 Billion by 2030," *PV Magazine*, February 18, 2021, https://www.pv-magazine.com/2021/02/18/global-hydrogen-project-pipeline-expected-to-exceed-300-billion-by-2030/.
15. Matthews and Blunt, "Green Hydrogen Plant in Saudi Desert."
16. Michelle Della Vigna, Zoe Stavrinou, and Alberto Gandolfi, "Carbonomics: Innovation, Deflation and Affordable De-carbonization," Goldman Sachs, October 13, 2020, https://www.goldmansachs.com/insights/pages/gs-research/carbonomics-innovation-deflation-and-affordable-de-carbonization/report.pdf.

8. CARBON REMOVAL

The quote in the chapter-opening image caption is taken from Myles McCormick, "Occidental Claims Green Push 'Does More than Tesla,'" *Financial Times*, January 18, 2021, https://www.ft.com/content/eb8236e0-abfc-4d82-b6ff-540d36c501e9.

1. Prof. Dr. Sabine Fuss and Prof. Jan Minx, "What the Paris Agreement Means," MCC: Common Economics Blog, March 8, 2018, https://blog.mcc-berlin.net/post/article/what-the-paris-agreement-means.html.
2. Vincent Gonzales, Alan Krupnik, and Lauren Dunlap, "Carbon Capture and Storage 101," Resources for the Future, May 6, 2020, https://www.rff.org/publications/explainers/carbon-capture-and-storage-101/.
3. Krysta Biniek, Kimberly Henderson, Matt Rogers, and Gregory Santoni, "Driving CO_2 Emissions to Zero (and Beyond) with Carbon Capture, Use, and Storage," McKinsey Quarterly, June 30, 2020, https://www.mckinsey.com/business-functions/sustainability/our-insights/driving-co2-emissions-to-zero-and-beyond-with-carbon-capture-use-and-storage.
4. Lawrence Irlam, "Global Costs of Carbon Capture and Storage," Global CCS Institute, June 2017, https://www.globalccsinstitute.com/archive/hub/publications/201688/global-ccs-cost-updatev4.pdf.
5. Biniek, Henderson, Rogers, and Santoni, "Driving CO_2 Emissions to Zero."
6. "The Tax Credit for Carbon Sequestration (Section 45Q)," Congressional Research Service, June 8, 2021, https://sgp.fas.org/crs/misc/IF11455.pdf.
7. "Valero and BlackRock Partner with Navigator to Announce Large-Scale Carbon Capture and Storage Project," BusinessWire, March 16, 2021, https://www.businesswire.com/news/home/20210316005599/en/.
8. K. G. Austin, J. S. Baker, B. L. Sohngen, et al., "The Economic Costs of Planting, Preserving, and Managing the World's Forests to Mitigate Climate Change," *Nature Communications* 11 (2020), https://www.nature.com/articles/s41467-020-19578-z.pdf.
9. Analysis by David Shukman, "Brazil's Amazon: Deforestation 'Surges to 12-Year High,'" BBC News, November 30, 2020, https://www.bbc.com/news/world-latin-america-55130304.
10. "Forests and Climate Change," IUCN, February 2021, https://www.iucn.org/resources/issues-briefs/forests-and-climate-change#:~:text=Around%2025%25%20of%20global%20emissions,from%20deforestation%20and%20forest%20degradation.
11. "REDD+ Reducing Emissions from Deforestation and Forest Degradation," Food and Agriculture Organization of the United Nations, October 6, 2020, http://www.fao.org/redd/news/detail/en/c/1309984/.

12. Jonathan Shieber, "As the Western US Burns, a Forest Carbon Capture Monitoring Service Nabs Cash from Amazon & Bill Gates-Backed Fund," TechCrunch+, September 17, 2020, https://techcrunch.com/2020/09/17/as-the-western-us-burns-a-forest-carbon-capture-monitoring-service-nabs-cash-from-amazon-bill-gates-backed-fund/?guccounter=1&guce_referrer=aHR0cHM6Ly93d3cuZ29vZ2xlLmNvbS8&guce_referrer_sig=AQAAAIqTYtdQa8nqPLgSLkW77KDzDxHdJ-ueNw7tqRydVU-muwCYGZ47SLKD58pFX1Buf1WvcA_BQgtn_b45EU7D_79486Pokck6zaLZpTNS_d5XoOd_y_u69isQGeEmZcdazzT1bKyq3vqM9bzX1c1-gMxKLFFBXTo66n7bGs1n2opg.
13. Ella Adlen and Cameron Hepburn, "10 Carbon Capture Methods Compared: Costs, Scalability, Permanence, Cleanness," EnergyPost.eu, November 11, 2019, https://energypost.eu/10-carbon-capture-methods-compared-costs-scalability-permanence-cleanness/.
14. Biniek, Henderson, Rogers and Santoni, "Driving CO_2 Emissions to Zero."
15. Sean Silcoff, "B.C.'s Carbon Engineering Secures $68-Million to Commercialize CO_2-Removal Technology," *Globe and Mail*, last modified March 22, 2019, https://www.theglobeandmail.com/business/article-bcs-carbon-engineering-secures-68-million-to-commercialize-co/.
16. The Chief Staff, "Carbon Engineering Doubles Capacity of Proposed U.S. Facility," The Squamish Chief, September 24, 2019, https://www.squamishchief.com/news/local-news/carbon-engineering-doubles-capacity-of-proposed-u-s-facility-1.23956265.
17. Jeff Tolefson, "Sucking Carbon Dioxide from Air Is Cheaper than Scientists Thought," News, *Nature*, June 7, 2018, https://www.nature.com/articles/d41586-018-05357-w.
18. J. Rogelj, D. Shindell, K. Jiang, et al., "Mitigation Pathways Compatible with 1.5°C in the Context of Sustainable Development," in *Global Warming of 1.5°C. An IPCC Special Report on the Impacts of Global Warming of 1.5°C Above Pre-industrial Levels and Related Global Greenhouse Gas Emission Pathways, in the Context of Strengthening the Global Response to the Threat of Climate Change, Sustainable Development, and Efforts to Eradicate Poverty*, ed. V. Masson-Delmotte, P. Zhai, H.-O. Pörtner, et al. (IPCC, 2018), https://www.ipcc.ch/sr15/chapter/chapter-2/.
19. "Elon Musk to Offer $100 Million Prize for 'Best' Carbon Capture Tech," Reuters, January 22, 2021, https://www.nbcnews.com/science/environment/elon-musk-offer-100-million-prize-best-carbon-capture-tech-rcna234.
20. "$100 Million Prize for Carbon Removal," XPrize, accessed February 5, 2022, https://www.xprize.org/prizes/elonmusk.

9. BETTER TOGETHER

The quote in the chapter-opening image caption is taken from "The Future of Energy: The End of the Oil Age," *The Economist*, October 25, 2003, https://www.economist.com/leaders/2003/10/23/the-end-of-the-oil-age.

1. Benoit Faucon and Summer Said, "Sheikh Yamani, Mastermind of Saudi Oil Supremacy, Dies at 90," *Wall Street Journal*, last modified February 23, 2021, https://www.wsj.com/articles/sheikh-yamani-who-led-saudi-arabias-rise-to-oil-supremacy-dies-at-90-11614066753.
2. Simon Evans, "Analysis: World Has Already Passed 'Peak Oil,' BP Figures Reveal," CarbonBrief, September 15, 2020, https://www.carbonbrief.org/analysis-world-has-already-passed-peak-oil-bp-figures-reveal.

3. Brad Plumer, "Electric Cars Are Coming, and Fast. Is the Nation's Grid Up to It?" *New York Times*, January 29, 2021, https://www.nytimes.com/2021/01/29/climate/gm-electric-cars-power-grid.html.
4. "Hydrogen Economy Outlook: Key Messages," BloombergNEF, March 30, 2020, https://data.bloomberglp.com/professional/sites/24/BNEF-Hydrogen-Economy-Outlook-Key-Messages-30-Mar-2020.pdf.
5. Simon Evans, "Direct CO_2 Capture Machines Could Use 'a Quarter of Global Energy' in 2100," CarbonBrief, July 22, 2019, https://www.carbonbrief.org/direct-co2-capture-machines-could-use-quarter-global-energy-in-2100.
6. "Table of Solutions," Project Drawdown, accessed February 5, 2022, https://drawdown.org/solutions/table-of-solutions.
7. "We will make electricity so cheap that only the rich will burn candles," Quote Investigator, accessed February 5, 2022, https://quoteinvestigator.com/2012/04/10/rich-burn-candles/#:~:text=Edison%20is%20reported%20to%20have,the%20rich%20will%20burn%20candles.%E2%80%9D.
8. "Net Zero by 2050 Scenario," IEA, last modified May 2021, https://www.iea.org/data-and-statistics/data-product/net-zero-by-2050-scenario#overview.
9. "Fast Facts on Transportation Greenhouse Gas Emissions," EPA, accessed February 5, 2022, https://www.epa.gov/greenvehicles/fast-facts-transportation-greenhouse-gas-emissions.
10. "Net Zero by 2050 Scenario," IEA.
11. William F. Lamb, Thomas Wiedmann, Julia Pongratz, et al., "A Review of Trends and Drivers of Greenhouse Gas Emissions by Sector from 1990 to 2018," *Environmental Research Letters* 16, no. 7 (2021): 073005, https://iopscience.iop.org/article/10.1088/1748-9326/abee4e.
12. Michelle Della Vigna, Zoe Stavrinou, and Alberto Gandolfi, "Carbonomics: Innovation, Deflation and Affordable De-carbonization," Goldman Sachs, October 13, 2020, https://www.goldmansachs.com/insights/pages/gs-research/carbonomics-innovation-deflation-and-affordable-de-carbonization/report.pdf.

10. RISK MITIGATION

The statistic about Hurricane Katrina in the chapter-opening image caption comes from Stephanie K. Jones, "Hurricane Katrina, the Numbers Tell Their Own Story," *Insurance Journal*, August 26, 2015, https://www.insurancejournal.com/news/southcentral/2015/08/26/379650.htm.

1. Munich Re, "Insurance gap: Extreme Weather Risks," website accessed February 24, 2022, https://www.munichre.com/en/risks/extreme-weather.html
2. Mark Carney, "Breaking the Tragedy of the Horizon—Climate Change and Financial Stability," BIS, September 29, 2015, https://www.bis.org/review/r151009a.pdf.
3. Carney, "Breaking the Tragedy."
4. Carney, "Breaking the Tragedy."
5. IPCC Sixth Assessment Report, https://www.ipcc.ch/assessment-report/ar6/.
6. Stephen Leahy, "'Off-the-Charts' Heat to Affect Millions in U.S. in Coming Decades," *National Geographic*, July 16, 2019, https://www.nationalgeographic.com/environment/article/extreme-heat-to-affect-millions-of-americans.

7. Jacob Fenston, "D.C. Averages a Week of 100-Degree Days. Climate Change Could Make That Two Months," WAMU, July 16, 2019, https://wamu.org/story/19/07/16/d-c-averages-a-week-of-100-degree-days-climate-change-could-make-that-two-months/.
8. Rebecca Lindsey, "Climate Change: Global Sea Level," NOAA/Climate.gov, last modified October 7, 2021, https://www.climate.gov/news-features/understanding-climate/climate-change-global-sea-level.
9. William Sweet and John Marra, "Understanding Climate: Billy Sweet and John Marra Explain Nuisance Floods," NOAA/Climate.gov, last modified July 9, 2021, https://www.climate.gov/news-features/understanding-climate/understanding-climate-billy-sweet-and-john-marra-explain.
10. Carney, "Breaking the Tragedy."
11. Jonathan Woetzel, Dickon Pinner, Hamid Samandari, et al., "Will Mortgages and Markets Stay Afloat in Florida?" McKinsey Global Institute, April 27, 2020, https://www.mckinsey.com/business-functions/sustainability/our-insights/will-mortgages-and-markets-stay-afloat-in-florida.
12. Woetzel, Pinner, Samandari, et al. "Will Mortgages and Markets."
13. Zillow Research, "Ocean at the Door: More than 386,000 Homes at Risk of Coastal Flooding by 2050," Zillow, November 13, 2018, https://www.zillow.com/research/ocean-at-the-door-21931/.
14. "What Is the Inevitable Policy Response?" PRI, accessed February 7, 2022, https://www.unpri.org/inevitable-policy-response/what-is-the-inevitable-policy-response/4787.article.
15. Nadja Popovich, Livia Albeck-Ripka, and Kendra Pierre-Louis, "The Trump Administration Rolled Back More than 100 Environmental Rules. Here's the Full List," *New York Times*, last modified January 20, 2021, https://www.nytimes.com/interactive/2020/climate/trump-environment-rollbacks-list.html.
16. "Executive Order on Climate-Related Financial Risk," White House Briefing Room, May 20, 2021, https://www.whitehouse.gov/briefing-room/presidential-actions/2021/05/20/executive-order-on-climate-related-financial-risk/.
17. "Final Report: Recommendations of the Task Force on Climate-Related Financial Disclosures," TCFD, June 2017, https://assets.bbhub.io/company/sites/60/2020/10/FINAL-2017-TCFD-Report-11052018.pdf.
18. "Global Investors Driving Business Transition," Climate Action 100+, accessed February 7, 2022, https://www.climateaction100.org/.
19. "Task Force on Climate-Related Financial Disclosures: 2020 Status Report," TCFD, October 2020, https://assets.bbhub.io/company/sites/60/2020/09/2020-TCFD_Status-Report.pdf.
20. Giulia Christianson and Ariel Pinchot, "BlackRock Is Getting Serious About Climate Change. Is This a Turning Point for Investors?" World Resources Institute, January 27, 2020, https://www.wri.org/insights/blackrock-getting-serious-about-climate-change-turning-point-investors.
21. "An Update on the ISSB at COP26," IFRS, accessed February 7, 2022, https://www.ifrs.org/news-and-events/news/2021/11/An-update-on-the-ISSB-at-COP26/.
22. https://www.sec.gov/news/press-release/2022-46.
23. Larry Fink's 2021 Letter to CEOs, BlackRock, https://www.blackrock.com/us/individual/2021-larry-fink-ceo-letter.

24. Philipp Krueger, Zacharias Sautner, and Laura T. Starks, "The Importance of Climate Risks for Institutional Investors," SSRN, November 11, 2019, https://papers.ssrn.com/sol3/papers.cfm?abstract_id=3235190.
25. "Climate Science and Investing: Integrating Climate Science and Investing," AllianceBernstein & Columbia Climate School: The Earth Institute, 2021, https://www.alliancebernstein.com/corporate/en/corporate-responsibility/environmental-stewardship/columbia-partnership.html.
26. "Investing Lessons from Climate School: Class of 2021," AllianceBernstein, May 27, 2021, https://www.alliancebernstein.com/corporate/en/insights/esg-in-action/esg-in-action-investing-lessons-from-climate-school-class-of-2021.html.
27. Peter H. Diamandis, "Problems Are Goldmines," Diamandis.com, August 30, 2015, https://www.diamandis.com/blog/problems-are-goldmines.

11. DIVESTMENT

The quote in the chapter-opening image caption is taken from Bill McKibben, "The Case for Fossil-Fuel Divestment: On the Road with the New Generation of College Activists Fighting for the Environment," *Rolling Stone*, February 22, 2013, https://www.rollingstone.com/politics/politics-news/the-case-for-fossil-fuel-divestment-100243/.

1. C. L. Brown and Omohundro Institute of Early American History & Culture, *Moral Capital: Foundations of British Abolitionism* (Chapel Hill: Omohundro Institute and University of North Carolina Press, 2006), 4.
2. Brycchan Carey and Geoffrey Gilbert Plank, eds., *Quakers and Abolition* (Champaign: University of Illinois Press, 2014), 30.
3. Carey and Plank, *Quakers and Abolition*, 3.
4. Adele Simmons, "Outside Opinion: Skeptics Were Wrong; South Africa Divestment Worked," *Chicago Tribune*, December 15, 2013, https://www.chicagotribune.com/business/ct-xpm-2013-12-15-ct-biz-1215-outside-opinion-20131215-story.html.
5. Simmons, "Outside Opinion."
6. Simmons, "Outside Opinion."
7. Simmons, "Outside Opinion."
8. Rebecca Leber, "Divestment Won't Hurt Big Oil, and That's OK," *New Republic*, May 20, 2015, https://newrepublic.com/article/121848/does-divestment-work.
9. McKibben, "Case for Fossil-Fuel Divestment."
10. McKibben, "Case for Fossil Fuel Divestment."
11. Hannah Ritchie and Max Roser, "CO_2 and Greenhouse Gas Emissions," Our World in Data, last modified August 2020, https://ourworldindata.org/co2-and-other-greenhouse-gas-emissions.
12. "Stanford to Divest from Coal Companies," Stanford Report, May 6, 2014, https://news.stanford.edu/news/2014/may/divest-coal-trustees-050714.html.
13. "Stanford to Divest from Coal Companies," Stanford Report.
14. Jeffrey Ball, "The Truth About Stanford's Coal Divestment," *New Republic*, May 22, 2014, https://newrepublic.com/article/117871/stanfords-coal-divestment-shows-environmental-hurdles-ahead.
15. Prof. Daniel R. Fischel, "Fossil Fuel Divestment: A Costly and Ineffective Investment Strategy," Compass Lexecon, accessed February 7, 2022, http://divestmentfacts.com/pdf/Fischel_Report.pdf.

16. Bradford Cornell, “The Divestment Penalty: Estimating the Costs of Fossil Fuel Divestment to Select University Endowments,” SSRN, September 3, 2015, https://papers.ssrn.com/sol3/papers.cfm?abstract_id=2655603.
17. “Investment Return of 12.3 Percent Brings Yale Endowment Value to $29.4 Billion,” Yale News, October 1, 2018, https://news.yale.edu/2018/10/01/investment-return-123-brings-yale-endowment-value-294-billion.
18. Amy Whyte, “Yale Activists Want Divestment. David Swensen Isn’t Budging,” Institutional Investor, February 21, 2020, https://www.institutionalinvestor.com/article/b1kftwb98pdn9q/Yale-Activists-Want-Divestment-David-Swensen-Isn-t-Budging.
19. Zach Schonfeld, “Stanford Pulls Its Coal Investments, but Why Haven’t Other Divestment Movements Succeeded?” *Newsweek*, May 9, 2014, https://www.newsweek.com/many-ways-college-administrations-have-resisted-fossil-fuel-divestment-movement-250409; Schonfeld, “Stanford Pulls Its Coal Investments”; Letter of Marc Fleurbaey, Chair, Princeton Sustainable Investment Initiative, Princeton University, May 1, 2015, https://cpucresources.princeton.edu/sites/cpucresources/files/reports/Report-on-the-Princeton-Sustainable-Investment-Initiative-proposal.pdf.
20. Alan Livsey, “Lex in Depth: The $900bn Cost of Stranded Energy Assets,” *Financial Times*, February 4, 2020, https://www.ft.com/content/95efca74-4299-11ea-a43a-c4b328d9061c.
21. Livsey, “Lex in Depth.”
22. C. McGlade and P. Ekins, “The Geographical Distribution of Fossil Fuels Unused when Limiting Global Warming to 2 °C,” *Nature* 517 (2015): 187–190, https://doi.org/10.1038/nature14016.
23. Livsey, “Lex in Depth.”
24. Alicia Steiger, “Mother Nature Is Not Calling for Divestment,” SLS, May 20, 2019, https://law.stanford.edu/2019/05/20/mother-nature-is-not-calling-for-divestment/.
25. Daniel R. Fischel, “The Feel-Good Folly of Fossil-Fuel Divestment,” Opinion, *Wall Street Journal*, February 9, 2015, https://www.wsj.com/articles/daniel-r-fischel-the-feel-good-folly-of-fossil-fuel-divestment-1423527484; Mike Gaworecki, “Fossil Fuel Industry Funds Study That Concludes Fossil Fuel Divestment Is a Bad Idea,” DeSmog, February 11, 2015, https://www.desmogblog.com/2015/02/11/fossil-fuel-industry-funds-study-concludes-fossil-fuel-divestment-bad-idea.
26. Arjan Trinks, Bert Scholtens, and Machiel Mulder, “Fossil Fuel Divestment and Portfolio Performance,” ResearchGate, April 2018, https://www.researchgate.net/publication/324140601_Fossil_Fuel_Divestment_and_Portfolio_Performance.
27. Jeremy Grantham, “The Mythical Peril of Divesting from Fossil Fuels,” Commentary, Grantham Research Institute, June 13, 2018, http://www.lse.ac.uk/GranthamInstitute/news/the-mythical-peril-of-divesting-from-fossil-fuels/.
28. “UC’s Investment Portfolios Fossil Free; Clean Energy Investments Top $1 Billion,” University of California Office of the President, May 19, 2020, https://www.universityofcalifornia.edu/press-room/uc-s-investment-portfolios-fossil-free-clean-energy-investments-top-1-billion.
29. “Letter from President Paxson: Brown’s Actions on Climate Change,” News, Brown University, March 4, 2020, https://www.brown.edu/news/2020-03-04/climate.
30. “Relevant Investment Policies,” Columbia Finance, accessed February 7, 2022, https://www.finance.columbia.edu/content/relevant-investment-policies.
31. Jasper G. Goodman and Kelsey J. Griffin, “Harvard Will Move to Divest Its Endowment from Fossil Fuels,” *Harvard Crimson*, September 10, 2021, https://www.thecrimson.com/article/2021/9/10/divest-declares-victory/.

32. "Fast Facts: Endowments," IES > NES, accessed February 7, 2022, https://nces.ed.gov/fastfacts/display.asp?id=73.
33. "Pension Funds: Total Financial Assets, Level," Economic Research: FRED Economic Data, last modified September 23, 2021, https://fred.stlouisfed.org/series/BOGZ1FL594090005Q.
34. "New York City to Divest Pension Funds of Fossil Fuels," United Nations Climate Change, January 11, 2018, https://unfccc.int/news/new-york-city-to-divest-pension-funds-of-fossil-fuels.
35. Anne Barnard, "New York's $226 Billion Pension Fund Is Dropping Fossil Fuel Stocks," *New York Times*, last modified August 11, 2021, https://www.nytimes.com/2020/12/09/nyregion/new-york-pension-fossil-fuels.html.
36. "Commitments," Global Fossil Fuels Divestment Commitment Database, accessed February 7, 2022, https://divestmentdatabase.org/.
37. Patrick Jenkins, "Energy's Stranded Assets Are a Cause of Financial Stability Concern," *Financial Times*, March 2, 2020, https://www.ft.com/content/17b54f60-5ba5-11ea-8033-fa40a0d65a98.
38. "Coal 2019: Analysis and Forecasts to2024," IEA, December 2019, https://www.iea.org/reports/coal-2019.
39. "Innovation, Deflation and Affordable De-carbonization," Goldman Sachs Research, October 13, 2020, https://publishing.gs.com/content/research/en/reports/2020/10/13/b6c26e3c-4556-41f9-81e3-c3b96bee5eb9.html.
40. Thomas Clarkson, *The History of the Rise, Progress, and Accomplishment of the Abolition of the African Slave-Trade by the British Parliament*, 2 vols. (London, 1808), 1:262.

12. ESG INVESTING

The quote in the chapter-opening image caption is taken from Kofi Annan, "Kofi Annan's Address to World Economic Forum in Davos," February 1, 1999, https://www.un.org/sg/en/content/sg/speeches/1999-02-01/kofi-annans-address-world-economic-forum-davos.

1. "2018 Global Sustainable Investment Review," Global Sustainable Investment Alliance, accessed February 7, 2022, https://www.ussif.org/files/GSIR_Review2018F.pdf.
2. *Who Cares Wins: Connecting Financial Markets to a Changing World*, The Global Compact, accessed February 7, 2022, https://www.unepfi.org/fileadmin/events/2004/stocks/who_cares_wins_global_compact_2004.pdf.
3. The author attended this meeting in 2005 and contributed to the *Who Cares Wins* report.
4. *Who Cares Wins*, The Global Compact.
5. "About the PRI," PRI, accessed February 7, 2022, https://www.unpri.org/pri/about-the-pri.
6. "About the PRI," PRI.
7. Lorenzo Saa, "PRI Milestone: 500 Asset Owner Members," Top 1000 Funds, January 30, 2020, https://www.top1000funds.com/2020/01/pri-milestone-500-asset-owner-members/.
8. Alice Ross, "Tackling Climate Change—An Investor's Guide", *Financial Times*, September 20, 2019, https://www.ft.com/content/fa7a4400-d940-11e9-8f9b-77216ebe1f17.

9. Sonal Mahida, "Fiduciary Duty Is Not an Obstacle to Addressing ESG," Intentional Endowments Network, The Crane Institute of Sustainability, accessed February 7, 2022, https://www.intentionalendowments.org/fiduciary_duty_is_not_an_obstacle_to_addressing_esg#_ftn4.
10. "Fact Sheet," U.S. Department of Labor, Employee Benefits Security Administration, October 13, 2021, https://www.dol.gov/sites/dolgov/files/EBSA/about-ebsa/our-activities/resource-center/fact-sheets/notice-of-proposed-rulemaking-on-prudence-and-loyalty-in-selecting-plan-investments-and-exercising-shareholder-rights.pdf.
11. Robert G. Eccles and Svetlana Klimenko, "The Investor Revolution," *Harvard Business Review*, May–June 2019.
12. "Create a 1st-Class GRI Standards Sustainability Report ASAP," FBRH Consultants, accessed February 7, 2022, https://fbrh.co.uk/en/80-percent-of-the-world%E2%80%99s-250-largest-companies-report-according-to-gri#:~:text=The%20Global%20Reporting%20Initiative%20(GRI)%20is%20the%20gold%20standard%20when,accordance%20with%20the%20GRI%20Standards.
13. Michael Cohn, "Former FASB Member Marc Siegel Joins SASB," *Accounting Today*, January 10, 2019, https://www.accountingtoday.com/news/former-fasb-board-member-marc-siegel-joins-sasb-to-set-sustainability-standards.
14. Gregory Unruh, David Kiron, Nina Kurschwitz, et al., "Investing for a Sustainable Future: Investors Care More About Sustainability than Many Executives Believe," *MIT Sloan Management Review*, May 11, 2016.
15. Vanessa Cuerel Burbano, "Social Responsibility Messages and Worker Wage Requirements: Field Experimental Evidence from Online Labor Marketplaces," ResearchGate, June 2016, https://www.researchgate.net/publication/304670971_Social_Responsibility_Messages_and_Worker_Wage_Requirements_Field_Experimental_Evidence_from_Online_Labor_Marketplaces.
16. BusinessWire, "Nielsen: 50 Percent of Global Consumers Surveyed Willing to Pay More for Goods, Services from Socially-Responsible Companies, up from 2011," August 6, 2013.
17. Smruti Kulkarni and Arnaud Lefebvre, "How Can Sustainability Enhance Your Value Proposition," The Nielsen Company, 2018, https://www.nielsen.com/wp-content/uploads/sites/3/2019/05/sustainable-innovation-report.pdf.
18. Michael J. Hiscox and J. Hainmueller, "Buying Green? Field Experimental Tests of Consumer Support for Environmentalism," Working Paper, Harvard University, last modified October 18, 2017, https://scholar.harvard.edu/hiscox/publications/buying-green-field-experimental-tests-consumer-support-environmentalism.
19. George Serafeim, "Integrated Reporting and Investor Clientele," *Journal of Applied Corporate Finance* 27, no. 2 (Spring 2015): 34–51, https://onlinelibrary.wiley.com/doi/10.1111/jacf.12116.
20. Ashish Lodh, "ESG and the Cost of Capital," MSCI Research, February 25, 2020, https://www.msci.com/www/blog-posts/esg-and-the-cost-of-capital/01726513589.
21. Amir Amel-Zadeh and George Serafeim, "Why and How Investors Use ESG Information: Evidence from a Global Survey," *Financial Analysts Journal* 74, no. 3 (2018): 87–103, https://www.tandfonline.com/doi/abs/10.2469/faj.v74.n3.2.
22. Tim Verheyden, Robert G. Eccles, and Andreas Feiner, "ESG for All? The Impact of ESG Screening on Return, Risk, and Diversification," *Journal of Applied Corporate Finance* 28, no. 2 (January 2016): 47–55, https://www.researchgate.net/publication/333244777_ESG_for_All_The_Impact_of_ESG_Screening_on_Return_Risk_and_Diversification.

23. "Sustainable Reality: Analyzing Risk and Returns of Sustainable Funds," Morgan Stanley, 2019, https://www.morganstanley.com/content/dam/msdotcom/ideas/sustainable-investing-offers-financial-performance-lowered-risk/Sustainable_Reality_Analyzing_Risk_and_Returns_of_Sustainable_Funds.pdf.
24. Alastair Marsh, "BlackRock Joins Allianz, Invesco Saying ESG Outperformed," Bloomberg Green, May 18, 2020, https://www.bloomberg.com/news/articles/2020-05-18/blackrock-joins-allianz-invesco-saying-esg-funds-outperformed?cmpid=BBD052020_GREENDAILY&utm_medium=email&utm_source=newsletter&utm_term=200520&utm_campaign=greendaily&sref=q3MO9qbb.
25. "Sustainable Reality," Morgan Stanley.
26. Ola Mahmoud and Julia Meyer, "The Anatomy of Sustainability," SSRN, May 1, 2020, https://ssrn.com/abstract=3597700.
27. "Global Sustainable Fund Flows Report," Morningstar, accessed February 7, 2022, https://www.morningstar.com/lp/global-esg-flows.
28. Mathieu Benhamou, Emily Chasan, and Saijel Kishan, "The Biggest ESG Funds Are Beating the Market," Bloomberg Green, January 29, 2020, https://www.bloomberg.com/graphics/2020-ten-funds-with-a-conscience/.
29. "ESG Integration and Analysis in the Americas," CFA Institute, 2018, https://www.cfainstitute.org/en/research/survey-reports/esg-integration-americas-survey-report.
30. Patrick Temple-West, "Impact Investing Creeps into the CLO Market," *Financial Times*, September 2, 2019, https://www.ft.com/content/14c8b0dc-cd5a-11e9-99a4-b5ded7a7fe3f.
31. Witold Henisz, Tim Koller, and Robin Nuttall, "Five Ways That ESG Creates Value," McKinsey Quarterly, November 2019, https://www.mckinsey.com/~/media/McKinsey/Business%20Functions/Strategy%20and%20Corporate%20Finance/Our%20Insights/Five%20ways%20that%20ESG%20creates%20value/Five-ways-that-ESG-creates-value.ashx.
32. Jess Shankelman, "Can Private Equity Giant Invest in Oil While Saving the Planet?" Bloomberg Green, June 24, 2020, https://www.bloomberg.com/news/articles/2020-06-24/can-a-private-equity-giant-invest-in-oil-while-saving-the-planet.
33. "Proxy Voting and Shareholder Engagement," BlackRock, accessed February 7, 2022, https://www.blackrock.com/corporate/literature/fact-sheet/blk-responsible-investment-faq-global.pdf.
34. Ross Kerber, "United States: Shareholder Support for U.S. Climate Measures Hits Nearly 50 Percent—Report," Reuters, September 15, 2021, https://www.reuters.com/world/us/shareholder-support-us-climate-measures-hits-nearly-50-report-2021-09-15/#:~:text=The%20paper%2C%20from%20Institutional%20Shareholder,two%20years%2C%20the%20report%20found.
35. Robert G. Eccles and Colin Mayer, "Can a Tiny Hedge Fund Push ExxonMobil Towards Sustainability?" *Harvard Business Review*, January 20, 2021, https://hbr.org/2021/01/can-a-tiny-hedge-fund-push-exxonmobile-towards-sustainability.
36. Saijel Kishan and Joe Carroll, "The Little Engine That Won an Environmental Victory Over Exxon," Bloomberg Businessweek, June 9, 2021, https://www.bloomberg.com/news/articles/2021-06-09/engine-no-1-proxy-campaign-against-exxon-xom-marks-win-for-esg-activists?sref=3rbSWFkc.
37. Alastair Marsh and Siajel Kishan, "Engine No. 1's Exxon Win Provides Boost for ESG Advocates," Bloomberg Green, May 27, 2021, https://www.bloomberg.com/news/articles/2021-05-27/engine-no-1-s-exxon-win-signals-turning-point-for-esg-investors?sref=3rbSWFkc.

38. Aaron Yoon and Soohon Kim, "Analyzing Active Mutual Fund Managers' Commitment to ESG: Evidence from the United Nations Principles for Responsible Investment," Northwestern Kellogg, 2020, https://www.kellogg.northwestern.edu/faculty/research/researchdetail?guid=c5358c04-9849-11ea-a76a-0242ac160003.
39. "Signatory Directory," PRI, accessed February 7, 2022, https://www.unpri.org/signatories/signatory-directory.
40. "What Is Powering the ESG Investment Surge?" Goldman Sachs, August 2, 2017, https://www.reuters.com/brandfeatures/goldman-sachs/what-is-powering-the-esg-investing-surge.
41. Robert G. Eccles and Svetlana Klimenko, "The Investor Revolution."
42. Shiva Rajgopal and Richard Foster, "ABCs of ESG," Breaking Views, Reuters, August 10, 2018, https://www.breakingviews.com/features/guest-view-esg-ratings-arent-reliable-enough/.
43. Steven Arons, "Deutsche Bank's DWS Slumps After U.S., Germany ESG Probe," Bloomberg, August 26, 2021, https://www.bloomberg.com/news/articles/2021-08-26/dws-shares-fall-after-u-s-opens-probe-on-sustainability-claims?sref=3rbSWFkc.
44. "About GRI," GRI, accessed February 7, 2022, https://www.globalreporting.org/information/about-gri/Pages/default.aspx.
45. Kelly Tang, "Indexology Blog: Carbon Emissions History of the S&P 500," S&P Dow Jones Indices, accessed February 7, 2022, https://www.indexologyblog.com/2018/01/31/carbon-emissions-history-of-the-sp-500-and-its-sectors/.
46. "S&P 500 Sales by Year," multpl, accessed February 7, 2022, https://www.multpl.com/s-p-500-sales/table/by-year.
47. Imogen Rose Smith, "David Blood and Al Gore Want to Reach the Next Generation," Institutional Investor, September 8, 2015, https://www.institutionalinvestor.com/article/b14z9wt9vk3ycy/david-blood-and-al-gore-want-to-reach-the-next-generation.
48. Owen Walker, "Al Gore: Sustainability Is History's Biggest Investment Opportunity," *Financial Times*, April 29, 2018, https://www.ft.com/content/1757dc40-486f-11e8-8ee8-cae73aab7ccb.
49. "2014 Social Enterprise Conference: Closing Keynote with Al Gore," Columbia Business School, The Social Enterprise Program, accessed February 7, 2022, https://www.youtube.com/watch?v=8aYJx2pfg34.
50. "Our Firm," Generation, https://www.generationim.com/our-firm/.

13. THEMATIC IMPACT INVESTING

The quote in the chapter-opening image caption is taken from "Impact Investing Goes Mainstream with DBL Partners' $400 Million Fund," MarketWatch, June 23, 2015, http://www.dblpartners.vc/2015/06/impact-investing-goes-mainstream-with-dbl-partners-400-million-fund/.

1. "7 Insights from Asset Owners on the Rise of Sustainable Investing," Institute for Sustainable Investing, Morgan Stanley, May 28, 2020, https://www.morganstanley.com/ideas/sustainability-investing-institutional-asset-owners.
2. Aneel G. Karnani, "Doing Well by Doing Good: The Grand Illusion," Ross School of Business Paper No. 1141, *California Management Review*, August 1, 2010, https://papers.ssrn.com/sol3/papers.cfm?abstract_id=1593009.
3. DBL Partners, "Bay Area Equity Fund 1," Palico, accessed February 7, 2022, https://www.palico.com/funds/bay-area-equity-fund-i/2a9b204c006a4aa996005875086ad99d.

4. DBL Partners, "Impact Investing Goes Mainstream with DBL Partners' $400 Million Fund," MarketWatch, June 23, 2015, http://www.dblpartners.vc/2015/06/impact-investing-goes-mainstream-with-dbl-partners-400-million-fund/; "SJF Ventures Closes Fourth Fund at $125 Million," SJF, December 15, 2016, https://sjfventures.com/sjf-ventures-closes-fourth-fund-at-125-million/.
5. Presentation by SJF in Columbia Business School Climate Finance course, February 2019.
6. Nancy E. Pfund and Lisa A. Hagerman, "Response to 'How Investors Can (and Can't) Create Social Value,'" Up for Debate, *Stanford Social Innovation Review*, December 8, 2016, https://ssir.org/up_for_debate/how_investors_can_and_cant_create_social_value/pfund_hagerman#; Jessica Matthews and David Sternlicht (Cambridge Associates), Amit Bouri, Abhilash Mudaliar, and Hannah Schiff (Global Impact Investing Network), "Introducing the Impact Investing Benchmark," Global Impact Investing Network, 2015, https://thegiin.org/assets/documents/pub/Introducing_the_Impact_Investing_Benchmark.pdf.
7. Abhilash Mudaliar, Rachel Bass, Hannah Dithrich, and Noshin Nova, "2019 Annual Impact Investor Survey," Global Impact Investing Network, June 19, 2019, https://thegiin.org/research/publication/impinv-survey-2019#charts.
8. "Committed to Lasting Impact," Bain Capital, accessed February 7, 2022, https://www.baincapital.com/about-us.
9. "Former Massachusetts Governor Deval L. Patrick Joins Bain Capital to Launch New Business Focused on Investments with Significant Social Impact," Bain Capital, April 13, 2015, https://www.baincapital.com/news/former-massachusetts-governor-deval-l-patrick-joins-bain-capital-launch-new-business-focused.
10. "Entrepreneurship: What We Know About Bain Capital's $390 Million Double Impact Fund," ImpactAlpha, July 19, 2017, https://impactalpha.com/what-we-know-about-bain-capitals-390-million-double-impact-fund-8dd4e0c90571/.
11. "TPG Launches Matrix Renewables with The Rise Fund's Acquisition of 1GW of Solar PV Projects from Trina Solar" [press release], The Rise Fund, July 1, 2020, https://therisefund.com/news/tpg-launches-matrix-renewables-rise-funds-acquisition-1gw-solar-pv-projects-trina-solar.
12. "Global Impact: Leveraging More than 40 Years of Experience, KKR Global Impact Launched in 2018 to Invest in Solutions-Oriented Businesses," KKR, accessed February 7, 2022, https://www.kkr.com/businesses/global-impact.
13. "Our Impact," SJF, accessed February 7, 2022, https://sjfventures.com/impact/.
14. "World Energy Investment 2019," IEA, 2019, https://www.iea.org/reports/world-energy-investment-2019/financing-and-funding-trends.
15. Jacob Kastrenakes, "Fisker Files for Chapter 11 Bankruptcy Protection Following Karma Electric Sports Car Flop," The Verge, November 23, 2013, https://www.theverge.com/2013/11/23/5137856/fisker-automotive-files-chapter-11-bankruptcy-hybrid-technology-sale.
16. Dr. Maximilian Holland, "Tesla Passes 1 Million EV Milestone & Model 3 Becomes All Time Best Seller," CleanTechnica, March 10, 2020, https://cleantechnica.com/2020/03/10/tesla-passes-1-million-ev-milestone-and-model-3-becomes-all-time-best-seller/.
17. Colin McKerracher, Aleksandra O'Donovan, Nick Albanese, et al., "Electric Vehicle Outlook 2021," BloombergNEF, accessed February 7, 2022, https://about.bnef.com/electric-vehicle-outlook/.

14. IMPACT FIRST INVESTING

The quote on the chapter-opening page is taken from Bill Gates, "A New Model for Investing in Energy Innovation," GatesNotes, December 12, 2016, https://www.gatesnotes.com/Energy/Breakthrough-Energy-Ventures.

1. Brian L. Trelstad, "Impact Investing: A Brief History," Harvard Business School Faculty & Research, December 2016, https://www.hbs.edu/faculty/Pages/item.aspx?num=55902.
2. "Muhammad Yunus: Banker to the Poor (Part I)," excerpted from Dr. Denise Ames's book: *Human Rights: Towards a Global Values System*, The Center for Global Awareness, posted on May 3, 2016, https://thecenterforglobalawareness.wordpress.com/2016/05/03/muhammad-yunus-banker-to-the-poor-part-i/.
3. Grameen Bank, accessed February 8, 2022, http://www.grameenbank.org.
4. "Acumen's Patient Capital Model Is a New Approach to Solving Poverty," Acumen, accessed February 8, 2022, https://acumen.org/about/patient-capital/.
5. "The Birth of Philanthrocapitalism," *The Economist*, February, 25, 2006, https://www.economist.com/special-report/2006/02/25/the-birth-of-philanthrocapitalism.
6. Saadia Madsbjerg, "Bringing Scale to the Impact Investing Industry," The Rockefeller Foundation, August 15, 2018, https://www.rockefellerfoundation.org/blog/bringing-scale-impact-investing-industry/?doing_wp_cron=1594994019.2125260829925537109375.
7. "Acumen Entrepreneurs Build Solutions to the Toughest Challenges Facing the Poor," Acumen, accessed February 8, 2022, https://acumen.org/investment/d-light.
8. Tayo Akinyemi, "The Delight of d.light Design," The Next Billion, accessed February 8, 2022, https://nextbillion.net/the-delight-of-d-light-design/.
9. "d.light design," Crunchbase, accessed February 8, 2022, https://www.crunchbase.com/organization/d-light-design#section-overview.
10. "Lighting the Way: Roadmaps to Exits in Off-Grid Energy," Acumen, 2019, https://acumen.org/wp-content/uploads/Acumen-Exits-Off-Grid-Energy-Report.pdf.
11. "Acumen Announces Nearly $70 Million Close of For-Profit Off-Grid Energy Fund Through Its Subsidiary Acumen Capital Partners," GlobeNewswire, April 17, 2019, https://www.globenewswire.com/news-release/2019/04/17/1805389/0/en/Acumen-Announces-Nearly-70-Million-Close-of-For-Profit-Off-Grid-Energy-Fund-through-its-subsidiary-Acumen-Capital-Partners.html.
12. "US Venture Capital Achieved Respectable Double-Digit Return in 2017," Cambridge Associates, October 15, 2018, https://www.globenewswire.com/news-release/2018/10/15/1621269/0/en/US-Venture-Capital-Achieved-Respectable-Double-Digit-Return-in-2017.html.
13. "100 Million Lives Illuminated," Acumen, accessed February 8, 2022, https://acumen.org/dlight100m/.
14. "Charitable Giving Statistics," National Philanthropic Trust, accessed February 8, 2022, https://www.nptrust.org/philanthropic-resources/charitable-giving-statistics/.
15. Kathleen Elkins, "Here's How Many People in America Qualify as Super Rich," CNBC, September 13, 2018, https://www.cnbc.com/2018/09/12/wealth-x-heres-how-many-people-in-america-qualify-as-super-rich.html.
16. "SEC Adopts Rule Under Dodd-Frank Act Defining 'Family Offices,'" SEC, June 22, 2011, https://www.sec.gov/news/press/2011/2011-134.htm.
17. "Acumen: The Ability to See the World as It Is, the Audacity to See the World as It Could Be," Acumen Partners, accessed February 8, 2022, https://acumen.org/wp-content/uploads/Acumen-Partner-One-Pager-Q4-2019.pdf.

18. Gates, "A New Model."
19. "Every Year, the World Adds 51 Billion Tons of Greenhouse Gases to the Atmosphere," Breakthrough Energy, accessed February 8, 2022, https://www.b-t.energy/ventures/.
20. Gates, "A New Model."
21. "Every Year, the World Adds," Breakthrough Energy.
22. Scott P. Burger, Fiona Murray, Sarah Kearney, and Liquian Ma, "The Investment Gap That Threatens the Planet," *Stanford Social Innovation Review* (Winter 2018), https://primecoalition.org/wp-content/uploads/2017/12/Winter_2018_the_investment_gap_that_threatens_the_planet.pdf?x48191.

15. RENEWABLE ENERGY PROJECTS

The quote in the epigraph is from Evelyn Chang, "Warren Buffett Says He's Got a 'Big Appetite' for a Solar or Wind Project," CNBC, May 26, 2017, http://www.cnbc.com/2017/05/06/warren-buffett-says-hes-got-a-big-appetite-for-a-solar-or-wind-project.html.

1. "Geospatial Data Science," NREL, accessed February 8, 2022, https://www.nrel.gov/gis/solar.html.
2. Brian Kennedy and Cary Lynne Thigpen, "More U.S. Homeowners Say They Are Considering Home Solar Panels," Pew Research Center, December 17, 2019, https://www.pewresearch.org/fact-tank/2019/12/17/more-u-s-homeowners-say-they-are-considering-home-solar-panels/.
3. "Project Overview," Samson Solar, accessed February 8, 2022, https://samsonsolarenergycenter.com/#overview.
4. "Solar Market Insight Report 2020 Year in Review," SEIA, March 16, 2021, https://www.seia.org/research-resources/solar-market-insight-report-2020-year-review.
5. "Lease Rates for Solar Farms: How Valuable Is My Land?" SolarLandLease, accessed February 8, 2022, https://www.solarlandlease.com/lease-rates-for-solar-farms-how-valuable-is-my-land#:~:text=The%20most%20commonly%2Dasked%20question,%242%2C000%20per%20acre%2C%20per%20year.
6. Benjamin Mow, "STAT FAQs Part 2: Lifetime of PV Panels," State, Local & Tribal Governments, NREL, April 23, 2018, https://www.nrel.gov/state-local-tribal/blog/posts/stat-faqs-part2-lifetime-of-pv-panels.html#:~:text=NREL%20research%20has%20shown%20that,rate%20of%200.5%25%20per%20year.
7. "Solar Power Purchase Agreements," SEIA, accessed February 8, 2022, https://www.seia.org/research-resources/solar-power-purchase-agreements#:~:text=A%20solar%20power%20purchase%20agreement%20(PPA)%20is%20a%20financial%20agreement,at%20little%20to%20no%20cost.
8. "Solar Investment Tax Credit (ITC)," SEIA, accessed February 8, 2022, https://www.seia.org/initiatives/solar-investment-tax-credit-itc.
9. DSIRE Insight Team, "States Expanding Renewable and Clean Energy Standards," DSIREinsight, September 25, 2020, https://www.dsireinsight.com/blog/2020/9/25/states-expanding-renewable-and-clean-energy-standards.
10. "Renewable Portfolio Standard," State of New Jersey, Department of Environmental Protection, Air Quality, Energy & Sustainability, Office of Policy and Economic Standards, last modified February 16, 2017, https://www.state.nj.us/dep/aqes/opea-renewable-portfolio.html.

11. James Chen, "Renewable Energy Certificate," Investopedia, last modified May 3, 2021, https://www.investopedia.com/terms/r/rec.asp.
12. Catherine Lane, "What Is an SREC? Solar Renewable Energy Credits Explained," Solar Reviews, last modified May 7, 2021, https://www.solarreviews.com/blog/what-is-an-srec-and-how-can-i-get-the-best-srec-prices.
13. "Historic Auction Prices," SRECTrade, accessed February 8, 2022, https://www.srectrade.com/auction.
14. Vikram Aggarwal, "What to Know About a Solar Panel Warranty," EnergySage, January 20, 2021, https://news.energysage.com/shopping-solar-panels-pay-attention-to-solar-panels-warranty/.
15. "How Long Do Solar Panels Last?" igsenergy, accessed February 8, 2022, https://www.igs.com/energy-resource-center/energy-101/how-long-do-solar-panels-last.
16. Karl-Erik Stromsa, "Fitch: Solar Projects Much More Reliable Performers than Wind Farms," GTM: A Wood Mackenzie Business, February 10, 2020, https://www.greentechmedia.com/articles/read/fitch-solar-projects-much-more-reliable-performers-than-wind-farms.
17. Emma Foehringer Merchant, "Is the Utility-Scale Solar Industry in a Finance Bubble?" GTM: A Wood Mackenzie Business, January 23, 2019, https://www.greentechmedia.com/articles/read/is-the-utility-scale-solar-industry-in-a-finance-bubble.
18. "U.S. Average Annual Wind Speed at 80 Meters," Office of Energy Efficiency & Renewable Energy: WindExchange, accessed February 8, 2022, https://windexchange.energy.gov/maps-data/319.
19. *New York State Wind Energy Guidebook*, NYSERDA, November 28, 2021, https://www.nyserda.ny.gov/windguidebook.
20. "Electric Power Monthly," EIA, accessed February 8, 2022, https://www.eia.gov/electricity/monthly/epm_table_grapher.php?t=epmt_6_07_b.
21. Electricity generated = power rating × capacity factor × hours operated. In this example, electricity generated = 5 MW × 35% × 24 hours/day × 365 days/year = 15,330 MWh/year.
22. Stromsa, "Fitch: Solar Projects."
23. "Areas of Industrial Wind Facilities," AWEO.org, accessed February 8, 2022, http://www.aweo.org/windarea.html.
24. Elizabeth Weise, "Wind Energy Gives American Farmers a New Crop to Sell in Tough Times," *USA Today*, last modified February 20, 2020, https://www.usatoday.com/story/news/nation/2020/02/16/wind-energy-can-help-american-farmers-earn-money-avoid-bankruptcy/4695670002/.
25. "Will the Cost to Produce Corn Decrease After 2022?" Illinois farmdoc, November 23, 2021, http://www.farmdoc.illinois.edu/manage/actual_projected_costs.pdf.
26. Jennifer Oldham, "Wind Is the New Corn for Struggling Farmers," Bloomberg Businessweek, October 6, 2016, https://www.bloomberg.com/news/articles/2016-10-06/wind-is-the-new-corn-for-struggling-farmers.
27. "Chapter 13: Power Purchase Agreement," Windustry, accessed February 8, 2022, https://www.windustry.org/community_wind_toolbox_13_power_purchase_agreement#:~:text=Length%20of%20the%20Agreement,-PPAs%20are%20long&text=The%20stated%20term%20of%20most,25%20years%20is%20not%20unusual.
28. "Q4 2020: Renewable Energy Deal Tracker," GreenBiz, accessed February 8, 2022, https://www.greenbiz.com/sites/default/files/2021-01/gbg_renewable_leaderboard%202020-Q4.pdf.

29. James Kobus, Ali Ibrahim Nasrallah, and Jim Guidera, "The Role of Corporate Renewable Purchase Power Agreements in Supporting US Wind and Solar Deployment," Columbia/SIPA, Center on Global Energy Policy, March 2021, https://www.energypolicy.columbia.edu/sites/default/files/pictures/PPA%20report,%20designed%20v4,%203.17.21.pdf.
30. Richard Bowers (principal contributor), "U.S. Wind Energy Production Tax Credit Extended Through 2021," Today in Energy, EIA, January 28, 2021, https://www.eia.gov/todayinenergy/detail.php?id=46576.
31. McDermott Will & Emery, "IRS Provides Relief for Offshore Wind and Federal Land Projects," JD Supra, January 7, 2021, https://www.jdsupra.com/legalnews/irs-provides-relief-for-offshore-wind-9892655/#:~:text=The%20offshore%20wind%20ITC%20is,waters%20of%20the%20United%20States.
32. Nate Chute, "What Percentage of Texas Energy Is Renewable? Breaking Down the State's Power Sources from Gas to Wind," *Austin American-Statesman*, last modified February 19, 2021, https://www.statesman.com/story/news/2021/02/17/texas-energy-wind-power-outage-natural-gas-renewable-green-new-deal/6780546002/.
33. Amanda Luhavalja, "Texas Renewable Energy Credit Markets Advance; Green-e Prices Back Off," S&P Global Market Intelligence, August 28, 2020, https://www.spglobal.com/marketintelligence/en/news-insights/latest-news-headlines/texas-renewable-energy-credit-markets-advance-green-e-prices-back-off-60107962#:~:text=Texas%20vintage%202020%20RECs%20posted,up%2C%20period%20is%20March%2031.
34. Tyler Hodge (principal contributor), "Wholesale U.S. Electricity Prices Were Generally Lower and Less Volatile in 2020 than 2019," Today in Energy, EIA, January 8, 2021, https://www.eia.gov/todayinenergy/detail.php?id=46396.
35. "About New York Green Bank," Green Bank Network, last modified June 2, 2020, https://greenbanknetwork.org/ny-green-bank/.
36. "Wind Turbine Reliability," Exponent, June 15, 2017, https://www.exponent.com/knowledge/alerts/2017/06/wind-turbine-reliability/?pageSize=NaN&pageNum=0&loadAllByPageSize=true.
37. "How Do Wind Turbines Survive Severe Storms?" Office of Energy Efficiency & Renewable Energy, accessed February 8, 2022, https://www.energy.gov/eere/articles/how-do-wind-turbines-survive-severe-storms.
38. Rochelle Toplensky, "Why Investors Have Learned to Love Wind and Solar Power," *Wall Street Journal*, June 6, 2020, https://www.wsj.com/articles/why-investors-have-learned-to-love-wind-and-solar-power-11594027941.
39. Luis Garcia, "Renewable Energy Investors Seek Returns in Project Development," Private Equity News, January 4, 2021, https://www.penews.com/articles/renewable-energy-investors-seek-returns-in-project-development-20210104.
40. "U.S. Storage Market Sets New Installation Record in Q3 2021," Wood Mackenzie, December 9, 2021, https://www.woodmac.com/industry/power-and-renewables/us-energy-storage-monitor/.
41. Eric Wesoff and William Driscoll, "How Does the US Retire 236 GW of Coal and 1,000 Gas Peaker Plants?" *PV Magazine*, September 18, 2020, https://pv-magazine-usa.com/2020/09/18/how-does-the-us-retire-236-gw-of-coal-and-1000-gas-peaker-plants/.
42. Michael J. Coren, "Solar Plus Batteries Aim to Retire Natural Gas Plants in 2019," *Quartz*, January 11, 2019, https://qz.com/1521660/solar-and-batteries-are-retiring-natural-gas-plants/.

43. "Energy Storage," California Energy Commission—Tracking Progress, last modified August 2018, https://www.energy.ca.gov/sites/default/files/2019-12/energy_storage_ada.pdf.
44. Robert Walton, "NextEra Inks 700 MW Wind + Solar + Battery Project, Largest in the US," Utility Dive, July 29, 2019, https://www.utilitydive.com/news/nextera-inks-700-mw-wind-solar-battery-project-largest-in-the-us/559693/.
45. Karl-Erik Stromsta, "Next Era Sees Little Threat to Wind and Solar from Fading Tax Credits," Greentech Media, July 24, 2019, https://www.greentechmedia.com/articles/read/nextera-sees-little-threat-to-wind-and-solar-from-fading-tax-credits.
46. Deanne Barrow, "Energy Storage: Warranties, Insurance and O&M Issues," Norton Rose Fulbright: Project Finance, June 19, 2019, https://www.projectfinance.law/publications/2019/june/energy-storage-warranties-insurance-and-om-issues/.
47. "Lazard's Levelized Cost of Storage Analysis—Version 6.0," Lazard, accessed February 8, 2022, https://www.lazard.com/media/451566/lazards-levelized-cost-of-storage-version-60-vf2.pdf.
48. Hodgson Russ LLP, "New York City Clears the Path for Permitting of Energy Storage Systems," JD Supra, December 17, 2020, https://www.jdsupra.com/legalnews/new-york-city-clears-the-path-for-52979/.
49. Paul Robson and Davide Bonomi, "Growing the Battery Storage Market 2020: Exploring Four Key Issues," from the Producers of the Energy Storage World Forum, Dufresne—Energy Storage World Forum, Dufresne Research Ltd., accessed February 8, 2022, https://energystorageforum.com/files/ESWF_Whitepaper_-_Growing_the_battery_storage_market.pdf.
50. "Lazard's Levelized Cost of Storage Analysis—Version 6.0," Lazard.
51. Darrell Proctor, "Distributed Energy: 'Best Is Yet to Come' for Energy Storage Technology," *Power*, March 1, 2021, https://www.powermag.com/best-is-yet-to-come-for-energy-storage-technology/.
52. Veronika Henze (contact), "Energy Storage Investments Boom as Battery Costs Halve in the Next Decade," BloombergNEF, July 31, 2019, https://about.bnef.com/blog/energy-storage-investments-boom-battery-costs-halve-next-decade/.
53. "Special Report: Global Renewables Performance Review (Solar and Wind Withstand Pandemic)," Fitch Ratings, March 15, 2021, https://www.fitchratings.com/research/infrastructure-project-finance/global-renewables-performance-review-solar-wind-withstand-pandemic-15-03-2021.
54. Veronika Henze (contact), "Energy Transition Investment Hit $500 Billion in 2020—for First Time," BloombergNEF, January 19, 2021, https://about.bnef.com/blog/energy-transition-investment-hit-500-billion-in-2020-for-first-time/.

16. REAL ESTATE

The quote in the chapter-opening caption is taken from Sarah Kaplan and Aaron Steckelberg, "Climate Solutions: Empire State of Green," *Washington Post*, May 27, 2020, https://www.washingtonpost.com/graphics/2020/climate-solutions/empire-state-building-emissions/.

1. Jonathan Shaw, "A Green Empire: How Anthony Malkin '84 Engineered the Largest "Green" Retrofit Ever," *Harvard Magazine*, March–April 2012, https://www.harvardmagazine.com/2012/03/a-green-empire.
2. Kaplan and Steckelberg, "Climate Solutions."

3. Shaw, "A Green Empire."
4. Kaplan and Steckelberg, "Climate Solutions."
5. "Sources of Greenhouse Gas Emissions," EPA, accessed February 8, 2022, https://www.epa.gov/ghgemissions/sources-greenhouse-gas-emissions.
6. "Retrofit Market Analysis," Urban Green, June 18, 2019, https://www.urbangreencouncil.org/sites/default/files/urban_green_retrofit_market_analysis.pdf.
7. "Empire State Building Retrofits Cut 10-Year Emissions by 40 Percent," The Energy Mix, June 2, 2020, https://theenergymix.com/2020/06/02/empire-state-building-retrofits-cut-10-year-emissions-by-40/.
8. Stefan Knupfer, "A Carbon Emission Reduction Toolkit for Global Cities," McKinsey Sustainability, June 5, 2019, https://www.mckinsey.com/business-functions/sustainability/our-insights/sustainability-blog/a-carbon-emission-reduction-toolkit-for-global-cities.
9. George Caraghiaur, "The Benefits of PACE Financing for Commercial Real Estate Companies," PACENation, May 2016, https://www.reit.com/sites/default/files/media/PDFs/ThebenefitsofPACEforCREFINAL.pdf.
10. Asaf Bernstein, Matthew Gustafson, and Ryan Lewis, "Disaster on the Horizon: The Price Effect of Sea Level Rise," *Journal of Financial Economics*, last updated July 26, 2018, https://papers.ssrn.com/sol3/papers.cfm?abstract_id=3073842.
11. Jonathan Woetzel, Dickon Pinner, Hamid Samandari, et al., "Will Mortgages and Markets Stay Afloat in Florida?" McKinsey Sustainability, April 27, 2020, https://www.mckinsey.com/business-functions/sustainability/our-insights/will-mortgages-and-markets-stay-afloat-in-florida.
12. "802,555 Homes at Risk of 10-Year Flood Inundation by 2050," Climate Central—Zillow Research, July 31, 2019, https://www.zillow.com/research/homes-at-risk-coastal-flooding-25040/.
13. A. Park Williams, John T. Abatzoglou, Alexander Gershunov, et al., "Observed Impacts of Anthropogenic Climate Change on Wildfire in California," *Earth's Future* 7, no. 8 (August 2019): 892–910, https://agupubs.onlinelibrary.wiley.com/doi/full/10.1029/2019EF001210.
14. James M. Vose and David L. Peterson (federal coordinating lead authors), "Chapter 6: Forests," in *Fourth National Climate Assessment*, (Washington, DC: U.S. Global Change Research Program, 2018), https://nca2018.globalchange.gov/chapter/6/.
15. Kate Mackenzie, "Lenders with the Best Climate Data Will Be in a Position to Discriminate," Bloomberg Green, June 26, 2020, https://www.bloomberg.com/news/articles/2020-06-26/lenders-with-the-best-climate-data-will-be-in-a-position-to-discriminate.
16. Christopher Flavelle, "Rising Seas Threaten an American Institution: The 30-Year Mortgage," *New York Times*, last modified March 2, 2021, https://www.nytimes.com/2020/06/19/climate/climate-seas-30-year-mortgage.html.
17. Christopher Flavelle, "As Wildfires Get Worse, Insurers Pull Back from Riskiest Areas," *New York Times*, August 20, 2019, https://www.nytimes.com/2019/08/20/climate/fire-insurance-renewal.html.
18. Jonathan Woetzel, Dickon Pinner, Hamid Samandari, et al., "Can Coastal Cities Turn the Tide on Rising Flood Risk?" McKinsey Sustainability, April 20, 2020, https://www.mckinsey.com/business-functions/sustainability/our-insights/can-coastal-cities-turn-the-tide-on-rising-flood-risk.

19. Anne Barnard, "The $119 Billion Sea Wall That Could Defend New York . . . or Not," *New York Times*, last modified August 21, 2021, https://www.nytimes.com/2020/01/17/nyregion/sea-wall-nyc.html.
20. Neal E. Robbins, "Deep Trouble: Can Venice Hold Back the Tide?" *The Guardian*, December 10, 2019, https://www.theguardian.com/environment/2019/dec/10/venice-floods-sea-level-rise-mose-project.
21. "Life's a Beach," Insight, Freddie Mac, April 26, 2016, http://www.freddiemac.com/research/insight/20160426_lifes_a_beach.page.
22. Jake Goodman, "Real Estate Powers Ahead on Decarbonisation," PRI, April 28, 2020, https://www.unpri.org/pri-blog/real-estate-powers-ahead-on-decarbonisation/5718.article.

17. FORESTRY AND AGRICULTURE

The quote on the chapter-opening page is taken from Gil Gullickson, "How Carbon May Become Another Crop for Farmers," *Successful Farming*, February 4, 2021, https://www.agriculture.com/farm-management/programs-and-policies/how-carbon-may-become-another-crop-for-farmers.

1. "The State of the World's Forests 2020," Food and Agriculture Organization of the UnitedNations, accessed February 8, 2022, http://www.fao.org/state-of-forests/en/#:~:text=The%20total%20forest%20area%20is,equally%20distributed%20around%20the%20globe.
2. Mikaela Weisse and Elizabeth Dow Goldman, "The World Lost a Belgium-Sized Area of Primary Rainforests Last Year," World Resources Institute, April 25, 2019, https://www.wri.org/blog/2019/04/world-lost-belgium-sized-area-primary-rainforests-last-year.
3. "Data Explorer," ClimateWatch, accessed February 8, 2022, https://www.climatewatchdata.org/data-explorer/historical-emissions?historical-emissions-data-sources=71&historical-emissions-gases=246&historical-emissions-regions=All%20Selected%2CWORLD&historical-emissions-sectors=All%20Selected&page=1&sort_col=country&sort_dir=DESC.
4. Kenneth Gillingham and James H. Stock, "The Cost of Reducing Greenhouse Gas Emissions," August 2, 2018, https://scholar.harvard.edu/files/stock/files/gillingham_stock_cost_080218_posted.pdf.
5. "Pathways to a Low-Carbon Economy: Version 2 of the Global Greenhouse Gas Abatement Cost Curve," McKinsey & Company, accessed February 8, 2022, https://www.mckinsey.com/~/media/McKinsey/Business%20Functions/Sustainability/Our%20Insights/Pathways%20to%20a%20low%20carbon%20economy/Pathways%20to%20a%20low%20carbon%20economy.pdf.
6. Raphael Calel, "Climate Change and Carbon Markets: A Panoramic History," Centre for Climate Change Economics and Policy Working Paper No. 62; Grantham Research Institute on Climate Change and the Environment Working Paper No. 52, July 2011, http://eprints.lse.ac.uk/37397/1/Climate_change_and_carbon_markets_a_panoramic_history(author).pdf.
7. L. Chestnut and D. M. Mills, "A Fresh Look at the Benefits and Costs of the US Acid Rain Program," *Journal of Environmental Management* 77 (2005): 252–66, https://cfpub.epa.gov/si/si_public_record_report.cfm?dirEntryID=139587.

8. William L. Anderegg, Anna T. Trugman, Grayson Badgley, et al., "Climate-Driven Risks to the Climate Mitigation Potential of Forests," *Science* 368, no. 6497 (June 19, 2020), https://science.sciencemag.org/content/368/6497/eaaz7005.
9. Ovidiu Csillik, Pramukta Kumar, Joseph Mascaro, et al., "Monitoring Tropical Forest Carbon Stocks and Emissions Using Planet Satellite Data," *Scientific Reports* 9 (2019): 17831, https://www.nature.com/articles/s41598-019-54386-6.
10. "Value of Carbon Market Update 2021," Carbon Credit Capital, accessed February 8, 2022, https://carboncreditcapital.com/value-of-carbon-market-update-2021-2/.
11. "Voluntary Carbon Markets Rocket in 2021, on Track to Break $1B for First Time," Ecosystem Marketplace, accessed February 8, 2022, https://www.ecosystemmarketplace.com/articles/press-release-voluntary-carbon-markets-rocket-in-2021-on-track-to-break-1b-for-first-time/.
12. "Cap-and-Trade Program," California Air Resources Board, accessed February 8, 2022, https://www.arb.ca.gov/cc/capandtrade/capandtrade.htm.
13. "ARB Offset Credit Issuance," California Air Resources Board, accessed February 26, 2022, https://ww2.arb.ca.gov/our-work/programs/compliance-offset-program/arb-offset-credit-issuance.
14. Jessica Bursztynsky, "Delta Air Lines CEO Announces the Carrier Will Go 'Fully Carbon Neutral' Next Month," CNBC, February 14, 2020, https://www.cnbc.com/2020/02/14/delta-air-lines-ceo-carrier-will-go-fully-carbon-neutral-next-month.html.
15. "Delta Spotlights Ambitious Carbon Neutrality Plan on Path to Zero-Impact Aviation This Earth Month," Delta News Hub, April 22, 2021, https://news.delta.com/delta-spotlights-ambitious-carbon-neutrality-plan-path-zero-impact-aviation-earth-month.
16. ICE Report Center, accessed February 8, 2022, https://www.theice.com/marketdata/reports/142.
17. Ryan Dezember, "Preserving Trees Becomes Big Business, Driven by Emissions Rules," *Wall Street Journal*, last modified August 24, 2020, https://www.wsj.com/articles/preserving-trees-becomes-big-business-driven-by-emissions-rules-11598202541.
18. Jean-Francois Bastin, Yelena Finegold, Danilo Mollicone, et al., "The Global Tree Restoration Potential," *Science* 365, no. 6448 (July 5, 2019): 76–79, https://science.sciencemag.org/content/365/6448/76.
19. McKinsey and Company, "Putting carbon markets to work on the path to net zero," October 28, 2021, https://www.mckinsey.com/business-functions/sustainability/our-insights/putting-carbon-markets-to-work-on-the-path-to-net-zero.
20. Weisse and Goldman, "The World Lost."
21. "Pathways to a Low-Carbon Economy, Version 2 of the Global Greenhouse Gas Abatement Cost Curve," McKinsey & Company, accessed February 8, 2022, https://www.mckinsey.com/~/media/McKinsey/Business%20Functions/Sustainability/Our%20Insights/Pathways%20to%20a%20low%20carbon%20economy/Pathways%20to%20a%20low%20carbon%20economy.pdf.
22. Jeff Goodell, "Why Planting Trees Won't Save Us," *Rolling Stone*, June 25, 2020, https://www.rollingstone.com/politics/politics-features/tree-planting-wont-stop-climate-crisis-1020500/.
23. Bastin, Finegold, Garcia, et al., "Global Tree Restoration Potential."
24. Robert Heilmayr, Cristian Echeverría, and Eric F. Lambin, "Impacts of Chilean Forest Subsidies on Forest Cover, Carbon and Biodiversity," *Nature Sustainability* 3 (June 22, 2020): 701–9, https://www.nature.com/articles/s41893-020-0547-0.

25. H. de Coninck, A. Revi, M. Babiker, et al., "Strengthening and Implementing the Global Response," in *Global Warming of 1.5°C. An IPCC Special Report on the Impacts of Global Warming of 1.5°C Above Pre-industrial Levels and Related Global Greenhouse Gas Emission Pathways, in the Context of Strengthening the Global Response to the Threat of Climate Change, Sustainable Development, and Efforts to Eradicate Poverty*, ed. V. Masson-Delmotte, P. Zhai, H.-O. Pörtner, et al. (IPCC, 2018), https://www.ipcc.ch/site/assets/uploads/sites/2/2019/05/SR15_Chapter4_Low_Res.pdf.
26. Dom Phillips, "Bolsonaro Declares 'the Amazon Is Ours' and Calls Deforestation Data 'Lies,'" *The Guardian*, July 19, 2019, https://www.theguardian.com/world/2019/jul/19/jair-bolsonaro-brazil-amazon-rainforest-deforestation.
27. "Negative Emissions Technologies and Reliable Sequestration," National Academies Press, accessed February 8, 2022, https://www.nap.edu/resource/25259/interactive/.
28. "What Is BECCS?" American University, School of International Service, accessed February 8, 2022, https://www.american.edu/sis/centers/carbon-removal/fact-sheet-bioenergy-with-carbon-capture-and-storage-beccs.cfm.
29. "BECCS and Negative Emissions," drax, accessed February 8, 2022, https://www.drax.com/about-us/our-projects/bioenergy-carbon-capture-use-and-storage-beccs/.
30. Claire O'Connor, "Soil: The Secret Weapon in the Fight Against Climate Change," NRDC, December 5, 2019, https://www.nrdc.org/experts/claire-oconnor/soil-secret-weapon-fight-against-climate-change.
31. Elizabeth Creech, "Saving Money, Time and Soil: The Economics of No-Till Farming," USDA, August 3, 2021, https://www.usda.gov/media/blog/2017/11/30/saving-money-time-and-soil-economics-no-till-farming.
32. Gullickson, "How Carbon May Become."
33. Karl Plume, "Cargill Launches U.S. Carbon Farming Program for 2022 Season", Reuters, September 16, 2021, https://www.reuters.com/business/sustainable-business/cargill-launches-us-carbon-farming-program-2022-season-2021-09-16/.
34. Christopher Blaufelder, Cindy Levy, Peter Mannion, and Dickon Pinner, "A Blueprint for Scaling Voluntary Carbon Markets to Meet the Climate Challenge," McKinsey Sustainability, January 29, 2021, https://www.mckinsey.com/business-functions/sustainability/our-insights/a-blueprint-for-scaling-voluntary-carbon-markets-to-meet-the-climate-challenge.

SECTION 5: INVESTING IN FINANCIAL ASSETS

1. Catherine Clifford, "Blackrock CEO Larry Fink: The Next 1,000 Billion-Dollar Start-ups Will Be in Climate Tech," CNBC, October 5, 2021, https://www.cnbc.com/2021/10/25/blackrock-ceo-larry-fink-next-1000-unicorns-will-be-in-climate-tech.html.

18. VENTURE CAPITAL

The quote in the chapter-opening image caption is taken from "As filed with the Securities and Exchange Commission on April 30, 2019," SEC, https://www.sec.gov/Archives/edgar/data/1655210/000162828019004984/beyondmeats-1a6.htm.
1. "Economic Opportunity Quotes," Brainy Quote, accessed February 9, 2022, https://www.brainyquote.com/topics/economic-opportunity-quotes.

2. Benjamin Gaddy, Varun Sivaram, and Francis O'Sullivan, "Venture Capital and Clean Tech: The Wrong Model for Clean Energy Innovation," MIT Energy Initiative, July 2016, http://energy.mit.edu/wp-content/uploads/2016/07/MITEI-WP-2016-06.pdf.
3. Sarah Perez, "Why Did Solyndra Fail So Spectacularly?" TechCrunch+, October 4, 2011, https://techcrunch.com/2011/10/04/why-did-solyndra-fail-so-spectacularly/.
4. Yuliya Chernova, "Clouds Overtake Solar-Panel Firm," *Wall Street Journal*, September 1, 2011, https://www.wsj.com/articles/SB10001424053111904583204576542573023275438.
5. "Has Israeli Firm Cracked Eclectic Car Angst?" BBC News, September 2, 2012, https://www.bbc.com/news/world-middle-east-19423835.
6. Devashree Saha and Mark Muro, "Cleantech Venture Capital: Continued Declines and Narrow Geography Limit Prospects," Brookings, May 16, 2017, https://www.brookings.edu/research/cleantech-venture-capital-continued-declines-and-narrow-geography-limit-prospects/.
7. AllAboutAlpha, "Surfing the Sustainable Investing L-Curve: A Noble Way to Lose Money?" Investing.com, August 27, 2013, https://www.investing.com/analysis/surfing-the-sustainable-investing-l-curve:-a-noble-way-to-lose-money-181136.
8. Saha and Muro, "Cleantech Venture Capital."
9. "Clean Tech Company Performance Statistics 2018 Q4," Cambridge Associates, accessed February 9, 2022, https://www.cambridgeassociates.com/benchmarks/clean-tech-company-performance-statistics-2018-q4/.
10. "How Much Feed Does It Take to Produce a Pound of Beef?" Sustainable Dish, November 3, 2016, https://sustainabledish.com/much-feed-take-produce-pound-beef/.
11. "As filed with the Securities and Exchange Commission," SEC.
12. P. J. Gerber, H. Steinfeld, B. Henderson, et al., "Tackling Climate Change Through Livestock—A Global Assessment of Emissions and Mitigation Opportunities," Food and Agriculture Organization of the United Nations (FAO), 2013, http://www.fao.org/3/i3437e/i3437e.pdf.
13. "Beyond Meat Burger Carbon Footprint & Environmental Impact," Consumer Ecology, https://consumerecology.com/beyond-meat-burger-carbon-footprint-environmental-impact.
14. Sigal Samuel, "The Many Places You Can Buy Beyond Meat and Impossible Foods, in One Chart," Vox, last modified January 15, 2020, https://www.vox.com/future-perfect/2019/10/10/20870872/where-to-buy-impossible-foods-beyond-meat.
15. "Beyond Meat," crunchbase, accessed February 9, 2022, https://www.crunchbase.com/organization/beyond-meat/company_financials.
16. Ameelia Lucas, "Beyond Meat Surges 163 Percent in the Best IPO So Far in 2019," CNBC, May 2, 2019, https://www.cnbc.com/2019/05/02/beyond-meat-ipo.html.
17. Katherine White, David J. Hardisty, and Rishad Habib, "Consumer Behavior: The Elusive Green Consumer," *Harvard Business Review*, July–August 2019, https://hbr.org/2019/07/the-elusive-green-consumer#.
18. John D. Stoll, "Beyond Meat's Pitch for More Customers: It's Not Just Good for the Planet, It's Also Good for You," *Wall Street Journal*, September 25, 2020, https://www.wsj.com/articles/beyond-meats-pitch-for-more-customers-its-not-just-good-for-the-planet-its-also-good-for-you-11601042671.
19. Celine Herweijer and Azeem Azhar, "The State of Climate Tech," PwC, September 9, 2020, https://www.pwc.com/gx/en/services/sustainability/assets/pwc-the-state-of-climate-tech-2020.pdf.

20. "Global Climate Tech Venture Capital Report—Full Year 2021," HolonIQ, 4 January 2022, https://www.holoniq.com/notes/global-climatetech-vc-report-full-year-2021/.
21. Emma Cox, Tariq Moussa, and Denise Chan, "The State of Climate Tech 2020," PwC, November 30, 2017, https://www.pwc.com/gx/en/services/sustainability/publications/state-of-climate-tech-2020.html.
22. Jason D. Rowley, "Kleiner Perkins Spinout Fund G2VP Raises $298 Million for Oversubscribed Cleantech Fund," crunchbase news, April 24, 2018, https://news.crunchbase.com/news/kleiner-perkins-spinout-fund-g2vp-closes-298-million-over subscribed-cleantech-fund/.
23. Dina Bass, "Microsoft to Invest $1 Billion in Carbon-Reduction Technology," Bloomberg Green, January 16, 2020, https://www.bloomberg.com/news/articles/2020-01-16/microsoft-to-invest-1-billion-in-carbon-reduction-technology.
24. Dana Mattioli, "Amazon to Launch $2 Billion Venture Capital Fund to Invest in Clean Energy," *Wall Street Journal*, last modified June 23, 2020, https://www.wsj.com/articles/amazon-to-launch-2-billion-venture-capital-fund-to-invest-in-clean-energy-11592910001.
25. Akshat Rathi, "Bill Gates-Led Fund Raises Another $1 Billion to Invest in Clean Tech," Bloomberg Green, January 19, 2021, https://www.bloomberg.com/news/articles/2021-01-19/bill-gates-led-fund-raises-another-1-billion-to-invest-in-clean-tech?sref=3rbSWFkc.
26. Cromwell Schubarth, "Kleiner Perkins Has a Hearty Helping of Beyond Meat's Whopper of an IPO," *Silicon Valley Business Journal*, May 2, 2019, https://www.bizjournals.com/sanjose/news/2019/05/02/beyond-meat-ipo-big-investors-kleiner-perkins.html.
27. "SJF Ventures II, L.P.," Impactyield, last modified May 13, 2020, https://impactyield.com/funds/sjf-ventures-ii-l-p.
28. Connie Loizos, "Chris Sacca's Lowercarbon Capital Has Raised $800 Million to "Keep Unf*cking the Planet," TechCrunch, August 12, 2021, https://techcrunch.com/2021/08/12/chris-saccas-lowercarbon-capital-has-raised-800-million-to-keep-unfcking-the-planet/.

19. PRIVATE EQUITY

The quote in the chapter-opening image caption is from Isobel Markham, "In Depth: Carlyle's Head of Impact on the Firm's Approach to Climate Change," Private Equity International, September 24, 2020, https://www.privateequityinternational.com/in-depth-carlyles-head-of-impact-on-the-firms-approach-to-climate-change/.
1. Alastair Marsh, "Carney Calls Net-Zero Greenhouse Gas Ambition 'Greatest Commercial Opportunity'," Bloomberg Green, November 9, 2020, https://www.bloomberg.com/news/articles/2020-11-09/carney-calls-net-zero-ambition-greatest-commercial-opportunity?sref=3rbSWFkc.
2. Miriam Gottfried, "Blackstone Sets Goal to Reduce Carbon Emissions," *Wall Street Journal*, September 29, 2020, https://www.wsj.com/articles/blackstone-sets-goal-to-reduce-carbon-emissions-11601377200.
3. "Hilton Commits to Cutting Environmental Footprint in Half and Doubling Social Impact Investment," Business Wire, May 23, 2018, https://www.businesswire.com/news/home/20180522006472/en/Hilton-Commits-to-Cutting-Environmental-Footprint-in-Half-and-Doubling-Social-Impact-Investment.

4. "Energy Transition Investment Hit $500 Billion in 2020—for First Time," BloombergNEF, January 19, 2021, https://about.bnef.com/blog/energy-transition-investment-hit-500-billion-in-2020-for-first-time/.
5. Karl-Erik Stromsta, "BlackRock Targets Storage with New Multibillion-Dollar Renewables Fund," GTM: A Wood Mackenzie Business, January 27, 2020, https://www.greentechmedia.com/articles/read/blackrock-targets-storage-with-new-multi-billion-dollar-renewables-fund.
6. Stromsta, "BlackRock Targets Storage."
7. Jason Kelly and Derek Decloet, "Brookfield Pursues $7.5 Billion Fund Devoted to Net-Zero Shift," Bloomberg Green, February 10, 2021, https://www.bloomberg.com/news/articles/2021-02-10/brookfield-pursues-7-5-billion-fund-devoted-to-net-zero-shift?sref=3rbSWFkc.
8. Brian DeChesare, "Infrastructure Private Equity: The Definitive Guide," Mergers & Inquisitions, accessed February 9, 2022, https://www.mergersandinquisitions.com/infrastructure-private-equity/.
9. Marie Baudette, "Blackstone CEO Says Businesses Must Address Climate Change," *Wall Street Journal*, January 22, 2020, https://www.wsj.com/articles/blackstone-ceo-says-businesses-must-address-climate-change-11579721175.
10. Markham, "In Depth: Carlyle's Head of Impact."
11. TRD Staff, "How Blackstone Became the World's Biggest Landlord," The Real Deal, February 18, 2020, https://therealdeal.com/2020/02/18/how-blackstone-became-the-worlds-biggest-commercial-landlord/.

20. PUBLIC EQUITIES

The quote in the chapter-opening image caption is taken from Jim Robo, "CEO Letter," Next Era Energy, accessed February 9, 2022, https://www.nexteraenergy.com/sustainability/overview/ceo-letter.html.

1. "Net Zero Scorecard," Energy&Climate Intelligence Unit, accessed February 9, 2022, https://eciu.net/netzerotracker.
2. "Fact Sheet: President Biden Sets 2030 Greenhouse Gas Pollution Reduction Target Aimed at Creating Good-Paying Union Jobs and Securing U.S. Leadership on Clean Energy Technologies," White House, April 22, 2021, https://www.whitehouse.gov/briefing-room/statements-releases/2021/04/22/fact-sheet-president-biden-sets-2030-greenhouse-gas-pollution-reduction-target-aimed-at-creating-good-paying-union-jobs-and-securing-u-s-leadership-on-clean-energy-technologies/.
3. Tom Murray, "Apple, Ford, McDonald's and Microsoft Among This Summer's Climate Leaders," EDF, August 10, 2020, https://www.edf.org/blog/2020/08/10/apple-ford-mcdonalds-and-microsoft-among-summers-climate-leaders.
4. "Microsoft Announces It Will Be Carbon Negative by 2030," Microsoft News Center, January 16, 2020, https://news.microsoft.com/2020/01/16/microsoft-announces-it-will-be-carbon-negative-by-2030/#:~:text=on%20Thursday%20announced%20an%20ambitious,it%20was%20founded%20in%201975.
5. "Ford Expands Climate Change Goals, Sets Global Target to Become Carbon Neutral by 2050," Ford Media Center, June 26, 2020, https://media.ford.com/content/fordmedia/feu/en/news/2020/06/25/Ford-Expands-Climate-Change-Goals.html.

6. Sammy Roth, "Which Power Companies Are the Worst Polluters?" *Los Angeles Times*, June 26, 2019, https://www.latimes.com/projects/la-fi-power-companies-ranked-climate-change/.
7. Jeff St. John, "The 5 Biggest US Utilities Committing to Zero Carbon Emissions by 2050," GTM: A Wood Mackenzie Business, September 16, 2020, https://www.greentechmedia.com/articles/read/the-5-biggest-u.s-utilities-committing-to-zero-carbon-emissions-by-mid-century.
8. "Ford Expands Climate Change Goals, Sets Global Target to Become Carbon Neutral by 2050: Annual Sustainability Report," Ford Media Center, June 24, 2020, https://media.ford.com/content/fordmedia/fna/us/en/news/2020/06/24/ford-expands-climate-change-goals.html.
9. "We Are Taking Energy Forward: The Path to Net Zero and a Sustainable Energy Future," Baker Hughes, https://www.bakerhughes.com/sites/bakerhughes/files/2021-01/Baker%20Hughes%20-%20The%20path%20to%20net-zero%20and%20a%20sustainable%20energy%20future.pdf.
10. Kevin Crowley and Akshat Rathi, "Occidental to Strip Carbon from the Air and Use It to Pump Crude Oil," Bloomberg Businessweek, January 13, 2021, https://www.bloomberg.com/news/articles/2021-01-13/occidental-oxy-wants-to-go-green-to-produce-more-oil?sref=3rbSWFkc.
11. Jens Burchardt, Michel Frédeau, Miranda Hadfield, et al., "Supply Chains as a Game-Changer in the Fight Against Climate Change," BCG, January 26, 2021, https://www.bcg.com/publications/2021/fighting-climate-change-with-supply-chain-decarbonization.
12. Simon Jessop, "Sustainable Business: Company Bosses Question Benefits of Net Zero Transition - Survey," Reuters, March 25, 2021, 2https://www.reuters.com/business/sustainable-business/company-bosses-question-benefits-net-zero-transition-survey-2021-03-25/.
13. The EU Emissions Trading System (EU ETS), European Commission, accessed February 9, 2022, https://ec.europa.eu/clima/system/files/2016-12/factsheet_ets_en.pdf.
14. Dan Murtaugh, "Energy & Science: China's Carbon Market to Grow to $25 Billion by 2030, Citi Says," Bloomberg Green, March 8, 2021, https://www.bloomberg.com/news/articles/2021-03-08/china-s-carbon-market-to-grow-to-25-billion-by-2030-citi-says?sref=3rbSWFkc.
15. "State and Trends of Carbon Pricing 2021," World Bank Group, May 2021, https://openknowledge.worldbank.org/handle/10986/35620.
16. "What Is the Inevitable Policy Response?" PRI, accessed February 9, 2022, https://www.unpri.org/inevitable-policy-response/what-is-the-inevitable-policy-response/4787.article.
17. Darren Woods, "Why ExxonMobil Supports Carbon Pricing," EnergyFactor, March 29, 2021, https://energyfactor.exxonmobil.com/perspectives/supports-carbon-pricing/.
18. "General Motors, American Company," *Encyclopedia Britannica*, accessed February 9, 2022, https://www.britannica.com/topic/General-Motors-Corporation.
19. Neal Boudette, "G.M. expects production to return to normal this year as a chip shortage eases," *New York Times*, February 1, 2022, https://www.nytimes.com/2022/02/01/business/gm-earnings.html.
20. William P. Barnett, "Innovation: Why You Don't Understand Disruption," Stanford Business, March 7, 2017, https://www.gsb.stanford.edu/insights/why-you-dont-understand-disruption.

21. The Investopedia Team, "SunEdison: A Wall Street Boom-and-Bust Story," Investopedia, last modified June 30, 2020, https://www.investopedia.com/investing/sunedison-classic-wall-street-boom-and-bust-story/.
22. Liz Hoffman, "Inside the Fall of SunEdison, Once a Darling of the Clean-Energy World," *Wall Street Journal*, last modified April 14, 2016, https://www.wsj.com/articles/inside-the-fall-of-sunedison-once-a-darling-of-the-clean-energy-world-1460656000.
23. Tim McDonnell, "How Wind and Solar Toppled Exxon from Its Place as America's Top Energy Company," Quartz, November 30, 2020, https://qz.com/1933992/how-nextera-energy-replaced-exxon-as-the-us-top-energy-company/.
24. "NextEra Energy Is Once Again Recognized as No. 1 in Its Industry on Fortune's List of 'World's Most Admired Companies,'" Cision, February 2, 2021, https://www.prnewswire.com/news-releases/nextera-energy-is-once-again-recognized-as-no-1-in-its-industry-on-fortunes-list-of-worlds-most-admired-companies-301219998.html.
25. N. Sönnichsen, "Market Value of Global Electric Utilities 2021," statista.com, June 11, 2021, https://www.statista.com/statistics/263424/the-largest-energy-utility-companies-worldwide-based-on-market-value/.
26. "Elon Musk Quotes," Brainy Quote, accessed February 9, 2022, https://www.brainyquote.com/quotes/elon_musk_567297.
27. Justina Lee, "Rob Arnott Warns of Big Market Delusion in Electric Vehicles," Bloomberg Green, March 10, 2021, https://www.bloomberg.com/news/articles/2021-03-10/rob-arnott-warns-of-big-market-delusion-in-electric-vehicles?cmpid=BBD031721_GREENDAILY&utm_medium=email&utm_source=newsletter&utm_term=210317&utm_campaign=greendaily&sref=3rbSWFkc.
28. Brian Walsh, "SPACs Are the Constructs VCs Need to Fund Clean Tech," Tech Crunch+, January 28, 2021, https://techcrunch.com/2021/01/28/spacs-are-the-construct-vcs-need-to-fund-cleantech/?guccounter=1&guce_referrer=aHR0cHM6Ly93d3cuYW5nZWxsaXN0LmNvbS9ibG9nL2luc2lkZS10aGUtdmVudHVyZS1jbGltYXRlLXRlY2gtYm9vbQ&guce_referrer_sig=AQAAAKE3mTGOOoTvwRJyZse1uegosEyrm48O-8Nc5n-1vCvy-uoTmpELJNbxiNko_oovxqNj4B56hat9afy6hsC8LCwywA306eBuCfCluFoHxtfjh794_UbuRVwI22j42Pt1ZOJNXocqfPWUeE-YyIoTVodaRtmIUiPV_QfnYMBcm6om.
29. Yuliya Chernova, "SPAC Demand to Draw VCs to Clean Tech," *Wall Street Journal*, January 22, 2021, https://www.wsj.com/articles/spac-demand-to-draw-vcs-to-clean-tech-11611311402.
30. "The Yield Created by the Yieldco," Akin Gump, August 26, 2014, https://www.akingump.com/en/experience/industries/energy/speaking-energy/the-yield-created-by-the-yieldco.html.
31. Linette Lopez, "Wall Street's Getting Crushed by a Form of Financial Engineering You've Probably Never Heard Of," Business Insider, December 2, 2015, https://www.businessinsider.com/what-is-a-yieldco-and-how-is-it-killing-wall-street-2015-11.

21. EQUITY FUNDS

The quote in the epigraph is from Alistair Marsh and Sam Potter, "BlackRock Scores Biggest-Ever ETF Launch with New ESG Fund," Bloomberg Green, last modified April 9, 2021, https://www.bloomberg.com/news/articles/2021-04-09/blackrock-scores-record-etf-launch-with-carbon-transition-fund?sref=3rbSWFkc.

21. EQUITY FUNDS

1. "Founding Network Partners," Net Zero Asset Managers Initiative, accessed February 9, 2022, https://www.netzeroassetmanagers.org/#.
2. "The Net Zero Asset Managers Initiative Grows to 87 Investors Managing $37 Trillion, with the World's Three Largest Asset Managers Now Committing to Net Zero Goal," Net Zero Asset Managers Initiative, April 20, 2021, https://www.netzeroassetmanagers.org/the-net-zero-asset-managers-initiative-grows-to-87-investors-managing-37-trillion-with-the-worlds-three-largest-asset-managers-now-committing-to-net-zero-goal.
3. Attracta Mooney, "Vanguard Pledges to Slash Emissions by 2030," *Financial Times*, March 29, 2021, https://www.ft.com/content/87becf56-a249-4133-a01b-1b4b3b604bd5.
4. Attracta Mooney, "Fund Managers with $9tn in Assets Set Net Zero Goal," *Financial Times*, December 11, 2020, https://www.ft.com/content/d77d5ecb-4439-4f6b-b509-fffa42c194db.
5. "Our 2021 Stewardship Expectations," BlackRock, accessed February 9, 2022, https://www.blackrock.com/corporate/literature/publication/our-2021-stewardship-expectations.pdf.
6. Alex Cheema-Fox, Bridget Realmuto LaPerla, David Turkington, et al., "Decarbonization Factors," *Journal of Impact and ESG Investing* (Fall 2021), https://globalmarkets.statestreet.com/research/portal/insights/article/7151616d-aa8c-459d-b61d-98661f936100.
7. Jon Hale, "Which Sustainable Funds Are Fossil-Fuel Free?" Morningstar, April 22, 2020, https://www.morningstar.com/insights/2020/04/22/which-sustainable-funds-are-fossil-fuel-free.
8. "Oxford Martin Principles for Climate-Conscious Investment," Oxford Martin School Briefing, February 2018, https://www.oxfordmartin.ox.ac.uk/downloads/briefings/Principles_For_Climate_Conscious_Investment_Feb2018.pdf.
9. Jon Hale, "Building a Low-Cost, Fossil-Fuel-Free ETF Portfolio," Morningstar, June 17, 2019, https://www.morningstar.com/articles/934609/building-a-low-cost-fossil-fuel-free-etf-portfolio.
10. "ETF Comparison Tool: SPY vs SPYX," ETF.com, accessed February 9, 2022, https://www.etf.com/etfanalytics/etf-comparison/SPY-vs-SPYX.
11. "S&P 500 Fossil Fuel Free Index—ETF Tracker," ETF Database, accessed February 9, 2022, https://etfdb.com/index/sp-500-fossil-fuel-free-index/.
12. BlackRock Global Executive Committee, "Net Zero: A Fiduciary Approach," BlackRock, accessed February 9, 2022, https://www.blackrock.com/corporate/investor-relations/blackrock-client-letter.
13. Mats Andersson, Patrick Bolton, and Frédéric Samama, "Perspectives: Hedging Climate Risk," *Financial Analysts Journal* 72, no. 3, https://www0.gsb.columbia.edu/faculty/pbolton/papers/faj.v72.n3.4.pdf.
14. Robert G. Eccles and Svetlana Klimenko, "Finance and Investing: The Investor Revolution," *Harvard Business Review*, May–June 2019, https://hbr.org/2019/05/the-investor-revolution.
15. Marsh and Potter, "BlackRock Scores Biggest-Ever ETF Launch."
16. "BlackRock U.S. Carbon Transition Readiness ETF," iShares by BlackRock, accessed February 9, 2022, https://www.ishares.com/us/products/318215/blackrock-u-s-carbon-transition-readiness-etf.
17. "Solar Industry Research Data," SEIA, accessed February 9, 2022, https://www.seia.org/solar-industry-research-data.

18. Aakash Arora, Nathan Niese, Elizabeth Dreyer, et al., "Why Electric Cars Can't Come Fast Enough," BCG, April 20, 2021, https://www.bcg.com/publications/2021/why-evs-need-to-accelerate-their-market-penetration.
19. "Energy Transition: The Future for Green Hydrogen," Wood Mackenzie, October 25, 2019, https://www.woodmac.com/news/editorial/the-future-for-green-hydrogen/.
20. "iShares Global Clean Energy ETF," iShares by BlackRock, accessed February 26, 2022, https://www.ishares.com/us/products/239738/ishares-global-clean-energy-etf#/.
21. "iShares Global Clean Energy ETF," iShares by BlackRock.
22. "Greenbacker Renewable Energy Company," Greenbacker Capital, accessed February 9, 2022, https://greenbackercapital.com/greenbacker-renewable-energy-company/.
23. "KRBN: KraneShares Global Carbon ETF," KraneShares, accessed February 9, 2022, https://kraneshares.com/krbn/.
24. "S&P GSCI Carbon Emission Allowances (EUA)," S&P Dow Jones Indices, accessed February 9, 2022, https://www.spglobal.com/spdji/en/indices/commodities/sp-gsci-carbon-emission-allowances-eua/#overview.
25. "KraneShares Global Carbon ETF," KraneShares, accessed February 9, 2022, https://kraneshares.com/resources/factsheet/2021_04_30_krbn_factsheet.pdf.

22. FIXED INCOME

The quote in the epigraph is from Will Feuer, "Apple CEO Tim Cook Says He's Taking on Climate Change and Needs Backup," CNBC, October 22, 2019, https://www.cnbc.com/2019/10/22/apple-ceo-tim-cook-accepts-ceres-conference-sustainability-award.html.

1. James Langton, "A Record Year for Global Debt Issuance," Advisor's Edge, January 2, 2020, https://www.advisor.ca/news/industry-news/a-record-year-for-global-debt-issuance/.
2. Sophie Yeo, "Where Climate Cash Is Flowing and Why It's Not Enough," News Feature, *Nature*, September 17, 2019, https://www.nature.com/articles/d41586-019-02712-3.
3. Katie Kolchin, Justyna Podziemska, and Ali Mostafa, "Capital Markets Fact Book, 2021," sifma, July 28, 2021, https://www.sifma.org/resources/research/fact-book/.
4. Thomas Wacker, Andrew Lee, and Michaela Seimen Howat, "Sustainable Investing: Education Primer: Green Bonds," UBS, August 1, 2018, https://www.ubs.com/content/dam/WealthManagementAmericas/cio-impact/si-green-bonds-1-aug-2018-1523746.pdf.
5. Shuang Liu, "Will China Finally Block 'Clean Coal' from Green Bonds Market?" World Resources Institute, July 29, 2020, https://www.wri.org/insights/will-china-finally-block-clean-coal-green-bonds-market.
6. "10 Years of Green Bonds: Creating the Blueprint for Sustainability Across Capital Markets," World Bank, March 18, 2019, https://www.worldbank.org/en/news/immersive-story/2019/03/18/10-years-of-green-bonds-creating-the-blueprint-for-sustainability-across-capital-markets.
7. Malcolm Baker, Daniel Bergstresser, George Serafeim, and Jeffrey Wurgler, "Financing the Response to Climate Change: The Pricing and Ownership of U.S. Green Bonds," NBER, October 2018, https://www.nber.org/papers/w25194.
8. Matt Wirz, "Why Going Green Saves Bond Borrowers Money," *Wall Street Journal*, December 17, 2020, https://www.wsj.com/articles/why-going-green-saves-bond-borrowers-money-11608201002.

9. David F. Larcker and Edward M. Watts, "Where's the Greenium?" *Journal of Accounting and Economics* 69, no. 2 (2020), https://econpapers.repec.org/article/eeejaecon/v_3a69_3ay_3a2020_3ai_3a2_3as0165410120300148.htm.
10. Caroline Flammer, "Green Bonds Benefit Companies, Investors, and the Planet," *Harvard Business Review*, November 22, 2018, https://hbr.org/2018/11/green-bonds-benefit-companies-investors-and-the-planet.
11. Christopher Martin, "Green Bonds Show Path to $1 Trillion Market for Climate," Bloomberg, June 26, 2014, https://www.bloomberg.com/news/articles/2014-06-26/green-bonds-show-path-to-1-trillion-market-for-climate?sref=3rbSWFkc.
12. Liam Jones, "$500bn Green Issuance 2021: Social and Sustainable Acceleration: Annual Green $1tn in Sight," Climate Bonds Initiative, January 31, 2022, https://www.climatebonds.net/2022/01/500bn-green-issuance-2021-social-and-sustainable-acceleration-annual-green-1tn-sight-market.
13. Stephen Warwick, "Apple Says 1.2 Gigawatts of Clean Energy Produced by 2020 Green Bond Projects," iMore/Chrome Enterprise, March 17, 2021, https://www.imore.com/apple-says-12-gigawatts-clean-energy-produced-2020-green-bond-projects.
14. Billy Nauman, "Analysts Expect as Much as $500bn of Green Bonds in Bumper 2021," *Financial Times*, January 4, 2021, https://www.ft.com/content/021329aa-b0bd-4183-8559-0f3260b73d62.
15. Kristin Broughton, "Companies Test a New Type of ESG Bond with Fewer Restrictions," *Wall Street Journal*, October 5, 2020, https://www.wsj.com/articles/companies-test-a-new-type-of-esg-bond-with-fewer-restrictions-11601890200.
16. Meghna Mehta, "Green Bonds Are Growing Bigger and Broader," GreenBiz, May 4, 2020, https://www.greenbiz.com/article/green-bonds-are-growing-bigger-and-broader.
17. SEIA Comms Team, "A Look Back at Solar Milestones of the 2010s," SEIA, January 3, 2020, https://www.seia.org/blog/2010s-solar-milestones.
18. Herman K. Trabish, "Why Solar Financing Is Moving from Leases to Loans," Utility Dive, August 17, 2015, https://www.utilitydive.com/news/why-solar-financing-is-moving-from-leases-to-loans/403678/.
19. Kat Friedrich, "Solar and Energy Efficiency Securitization Emerge," Renewable Energy World, November 18, 2013, https://www.renewableenergyworld.com/2013/11/18/solar-and-energy-efficiency-securitization-emerge/#gref.
20. Mike Mendelsohn, "Raising Capital in Very Large Chunks: The Rise of Solar Securitization," *PV Magazine*, November 16, 2018, https://pv-magazine-usa.com/2018/11/16/raising-capital-in-very-large-chunks-the-rise-of-solar-securitization/.
21. Julia Pyper, "Solar Loans Emerge as the Dominant Residential Financing Product," GTM: A Wood Mackenzie Business, November 14, 2018, https://www.greentechmedia.com/articles/read/solar-loans-are-now-the-dominant-financing-product#gs.4fo6lu.
22. Julian Spector, "It's Official: Solar Securitizations Pass $1 Billion in 2017," GTM: A Wood Mackenzie Business, October 30, 2017, https://www.greentechmedia.com/articles/read/solar-securitizations-expected-to-pass-1-billion-in-2017.
23. "U.S. Residential Solar ABS 101," Project Bond Focus—July 2020: U.S. Residential Solar ABS 101, https://www.ca-cib.com/sites/default/files/2020-10/Project%20Bond%20Focus%20-%20Solar%20ABS%202020%20VF.pdf.
24. Julian Spector, "Mosaic Will Sell $300 Million Worth of Solar Loans to Goldman Sachs," GTM: A Wood Mackenzie Business, September 13, 2017, https://www.greentechmedia.com/articles/read/mosaic-will-sell-300-million-of-solar-loans-to-goldman-sachs.

25. Michael Kohick, "Muni Bond Defaults Remain Rare Through 2019," ETF Trends, September 19, 2020, https://www.etftrends.com/tactical-allocation-channel/muni-bond-defaults-remain-rare-through-2019/.
26. "Annual Report on Nationally Recognized Statistical Rating Organizations," SEC, January 2020, https://www.sec.gov/files/2019-annual-report-on-nrsros.pdf.
27. "Moody's Acquires Majority Stake in Four Twenty Seven Inc.," Moody's, July 24, 2019, https://ir.moodys.com/press-releases/news-details/2019/Moodys-Acquires-Majority-Stake-in-Four-Twenty-Seven-Inc-a-Leader-in-Climate-Data-and-Risk-Analysis/default.aspx.
28. Billy Nauman, "Municipal Bond Issuers Face Steeper Borrowing Costs from Climate Change," *Financial Times*, January 7, 2020, https://www.ft.com/content/6794c3d2-1d7d-11ea-9186-7348c2f183af.
29. Danielle Moran, "Muni Bonds Contain New Fine Print: Beware of Climate Change," Bloomberg Businessweek, November 5, 2019, https://www.bloomberg.com/news/articles/2019-11-05/how-serious-is-the-climate-change-risk-ask-a-banker?sref=3rbSWFkc.
30. Nauman, "Municipal Bond Issuers."
31. "Quantifying Wildfire Risk to Municipal Debt in California," risQ/MMA, accessed February 10, 2022, https://www.risq.io/wp-content/uploads/2019/03/risQ_MMA_QuantifyingWildfireRiskToMunicipalDebtInCalifornia-2.pdf.
32. Paul S. Goldsmith-Pinkham, Matthew Gustafson, Ryan Lewis, and Michael Schwert, "Sea Level Rise Exposure and Municipal Bond Yields," SSRN, October 6, 2021, https://papers.ssrn.com/sol3/papers.cfm?abstract_id=3478364.
33. Caroline Cournoyer, "Massachusetts Uses Popularity of Environmental Stewardship to Pad Its Bottom Line," Governing, June 26, 2013, https://www.governing.com/archive/gov-massachusetts-green-bonds-a-first.html.
34. Danielle Moran, "Biden Spending Plan Seen Jolting Muni Green Bond Sales to Record," Bloomberg Green, April 20, 2021, https://www.bloomberg.com/news/articles/2021-04-20/biden-spending-plan-seen-jolting-muni-green-bond-sales-to-record?sref=3rbSWFkc.
35. "Franklin Templeton Launches Franklin Municipal Green Bond Fund for US Investors," BusinessWire, August 4, 2020, https://www.businesswire.com/news/home/20200804005632/en/Franklin-Templeton-Launches-Franklin-Municipal-Green-Bond-Fund-for-US-Investors.

23. THE INVESTOR'S DILEMMA

1. "Humans Wired to Respond to Short-Term Problems," *Talk of the Nation*, NPR, July 3, 2006, https://www.npr.org/templates/story/story.php?storyId=5530483.
2. Greg Harman, "Your Brain on Climate Change: Why the Threat Produces Apathy, Not Action," *The Guardian*, November 10, 2014, https://www.theguardian.com/sustainable-business/2014/nov/10/brain-climate-change-science-psychology-environment-elections.
3. "Global Energy Review: CO_2 Emissions in 2021," IEA, Paris 2021, https://www.iea.org/reports/global-energy-review-2021/co2-emissions.
4. Alan Buis, "The Atmosphere: Getting a Handle on Carbon Dioxide," NASA: Global Climate Change, October 29, 2019, https://climate.nasa.gov/news/2915/the-atmosphere-getting-a-handle-on-carbon-dioxide.

5. John Kemp, "Climate Change Targets Are Slipping Out of Reach," Reuters, April 16, 2019, https://www.reuters.com/article/energy-climatechange-kemp/column-climate-change-targets-are-slipping-out-of-reach-kemp-idUSL5N21Y4A0.
6. "Summary for Policymakers," *Synthesis Report*, IPCC, accessed February 10, 2022, http://ar5-syr.ipcc.ch/topic_summary.php.
7. "Atmospheric CO_2 Data," Scripps CO2 Program, accessed February 10, 2022, https://scrippsco2.ucsd.edu/data/atmospheric_co2/primary_mlo_co2_record.html.
8. Michelle Della Vigna, Zoe Stavrinou, and Alberto Gandolfi, "Carbonomics: Innovation, Deflation and Affordable De-carbonization," Goldman Sachs, October 13, 2020, https://www.goldmansachs.com/insights/pages/gs-research/carbonomics-innovation-deflation-and-affordable-de-carbonization/report.pdf.
9. "Special Report: Global Warming of 1.5°C. Summary for Policymakers," IPCC, accessed February 10, 2022, https://www.ipcc.ch/sr15/chapter/spm/.
10. Allan Myles, Mustafa Babiker, Yang Chen, et al., "Summary for Policymakers," in *Global Warming of 1.5°C. An IPCC Special Report on the Impacts of Global Warming of 1.5°C Above Pre-industrial Levels and Related Global Greenhouse Gas Emission Pathways, in the Context of Strengthening the Global Response to the Threat of Climate Change, Sustainable Development, and Efforts to Eradicate Poverty*, ed. V. Masson-Delmotte, P. Zhai, H.-O. Pörtner, et al. (IPCC, 2018), https://www.ipcc.ch/site/assets/uploads/sites/2/2019/05/SR15_SPM_version_report_HR.pdf.
11. Joby Warrick and Chris Mooney, "Health & Science: Effects of Climate Change 'Irreversible,' U.N. Panel Warns in Report," *Washington Post*, November 2, 2014, https://www.washingtonpost.com/national/health-science/effects-of-climate-change-irreversible-un-panel-warns-in-report/2014/11/01/2d49aeec-6142-11e4-8b9e-2ccdac31a031_story.html?utm_term=.41a0c1bb1cff.

24. BEST PRACTICES

1. Nancy Weil, "The Quotable Bill Gates," ABC News, June 23, 2008, https://abcnews.go.com/Technology/PCWorld/story?id=5214635.
2. Haley Walker, "Recapping on BP's Long History of Greenwashing," Greenpeace, May 21, 2010, https://www.greenpeace.org/usa/recapping-on-bps-long-history-of-greenwashing/.
3. Kate Mackenzie, "A Legacy of Greenwashing Haunts BP's End-of-Oil Vision," Bloomberg Green, September 18, 2020, https://www.bloomberg.com/news/articles/2020-09-18/a-legacy-of-greenwashing-haunts-bp-s-end-of-oil-vision?sref=3rbSWFkc.
4. Jillian Ambrose, "BP Enjoys Share Bounce After Unveiling Plans to Shift Away from Fossil Fuels," *The Guardian*, August 4, 2020, https://www.theguardian.com/business/2020/aug/04/bp-dividend-covid-record-loss-energy-oil-gas.
5. Emily Bobrow, "Environment: Fight Climate Change with Behavior Change," Behavioral Scientist, October 16, 2018, https://behavioralscientist.org/fight-climate-change-with-behavior-change/.
6. Sigal Samuel, "The Many Places You Can Buy Beyond Meat and Impossible Foods, in One Chart," Vox, last modified January 15, 2020, https://www.vox.com/future-perfect/2019/10/10/20870872/where-to-buy-impossible-foods-beyond-meat.
7. Mark Carney, "Resolving the Climate Paradox," BIS, September 22, 2016, https://www.bis.org/review/r160926h.pdf.

25. WHY INVESTMENT MATTERS

"A Pale Blue Dot," The Planetary Society, accessed February 10, 2022, https://www.planetary.org/worlds/pale-blue-dot.

1. "Climate Change 2021: The Physical Science Basis," IPCC, accessed February 10, 2022, https://www.ipcc.ch/report/ar6/wg1/downloads/report/IPCC_AR6_WGI_SPM_final.pdf.
2. Scott A. Kulp and Benjamin H. Strauss, "New Elevation Data Triple Estimates of Global Vulnerability to Sea-Level Rise and Coastal Flooding," *Nature Communications* 10 (2019): article 4844, https://www.nature.com/articles/s41467-019-12808-z.
3. "Climate Change 2021," IPCC.
4. Jennifer Elks, "Havas: 'Smarter' Consumers Will Significantly Alter Economic Models and the Role of Brands," Sustainable Brands, May 14, 2014, https://sustainablebrands.com/read/defining-the-next-economy/havas-smarter-consumers-will-significantly-alter-economic-models-and-the-role-of-brands.
5. Maxine Joselow, "Quitting Burgers and Planes Won't Stop Warming, Experts Say," E&E News Climate Wire, December 6, 2019, https://www.eenews.net/articles/quitting-burgers-and-planes-wont-stop-warming-experts-say/.
6. Seth H. Werfel, "Household Behaviour Crowds Out Support for Climate Change Policy When Sufficient Progress Is Perceived," *Nature Climate Change* 7 (2017): 512–15, https://www.nature.com/articles/nclimate3316.
7. "Global Ideas: Climate Crisis: Is It Time to Ditch Economic Growth?" DW, accessed February 10, 2022, https://www.dw.com/en/climatechange-emissions-fossilfuels-gdp-economy-renewables/a-55089013.
8. Hannah Ritchie, "Who Has Contributed Most to Global CO_2 Emissions?" Our World in Data, October 1, 2019, https://ourworldindata.org/contributed-most-global-co2.
9. "Global Energy Review 2021: CO_2 Emissions," IEA, accessed February 10, 2022, https://www.iea.org/reports/global-energy-review-2021/co2-emissions.
10. Damian Carrington, "Want to Fight Climate Change? Have Fewer Children," *The Guardian*, July 12, 2017, https://www.theguardian.com/environment/2017/jul/12/want-to-fight-climate-change-have-fewer-children.
11. Carrington, "Want to Fight Climate Change?"
12. Lydia Denworth, "Children Change Their Parents' Minds About Climate Change," *Scientific American*, May 6, 2019, https://www.scientificamerican.com/article/children-change-their-parents-minds-about-climate-change/.
13. Abrahm Lustgarten, "How Russia Wins the Climate Crisis," *New York Times*, December 20, 2020, https://www.nytimes.com/interactive/2020/12/16/magazine/russia-climate-migration-crisis.html.
14. "Climate Finance Markets and the Real Economy," sifma, December 2020, https://www.sifma.org/wp-content/uploads/2020/12/Climate-Finance-Markets-and-the-Real-Economy.pdf.
15. "Climate Finance Markets," sifma.
16. "Climate Finance Markets," sifma.
17. "Bond Market Size," ICMA, accessed February 10, 2022, https://www.icmagroup.org/Regulatory-Policy-and-Market-Practice/Secondary-Markets/bond-market-size/.
18. Matthew Toole, "Global Capital Markets Answer 2020's Distress Call," Refinitiv, January 21, 2021, https://www.refinitiv.com/perspectives/market-insights/global-capital-markets-answer-2020s-distress-call/.

19. Ewa Skornas and Elisabeth Bautista Suarez, "2021 Global Private Equity Outlook," S&P Market Intelligence, March 2, 2021, https://www.spglobal.com/marketintelligence/en/news-insights/research/2021-global-private-equity-outlook.
20. "Indicator: Greenhouse Gas Emissions," Umwelt Bundesamt, July 22, 2021, https://www.umweltbundesamt.de/en/data/environmental-indicators/indicator-greenhouse-gas-emissions#at-a-glance.
21. "Germany—Gross Domestic Product per Capita in Constant Prices of 2010," Knoema, accessed February 10, 2022, https://knoema.com/atlas/Germany/topics/Economy/National-Accounts-Gross-Domestic-Product/Real-GDP-per-capita.
22. Julian Wettengel, "Renewables Produce More Power than Fossil Fuels in Germany for First Time," Clean Energy Wire, January 4, 2021, https://www.cleanenergywire.org/news/renewables-produce-more-power-fossil-fuels-germany-first-time.
23. Sören Amelang, "How Much Does Germany's Energy Transition Cost?" Clean Energy Wire, June 1, 2018, https://www.cleanenergywire.org/factsheets/how-much-does-germanys-energy-transition-cost.
24. Amelang, "How Much?"
25. Sophie Yeo and Simon Evans, "The 35 Countries Cutting the Link Between Economic Growth and Emissions," CarbonBrief, April 5, 2016, https://www.carbonbrief.org/the-35-countries-cutting-the-link-between-economic-growth-and-emissions.
26. Zeke Hausfather, "Absolute Decoupling of Economic Growth and Emissions in 32 Countries," Breakthrough Institute, April 6, 2021, https://thebreakthrough.org/issues/energy/absolute-decoupling-of-economic-growth-and-emissions-in-32-countries.
27. Zeke Hausfather, "Analysis: Why US Carbon Emissions Have Fallen 14 Percent Since 2005," CarbonBrief, August 15, 2017, https://www.carbonbrief.org/analysis-why-us-carbon-emissions-have-fallen-14-since-2005.
28. "Mutual Funds, Past Performance," Investor.gov, U.S. Securities and Exchange Commission, accessed February 10, 2022, https://www.investor.gov/introduction-investing/investing-basics/glossary/mutual-funds-past-performance.
29. National Academies of Sciences, Engineering, and Medicine, *Accelerating Decarbonization of the U.S. Energy System* (Washington, DC: National Academies Press, 2021), https://www.nap.edu/catalog/25932/accelerating-decarbonization-of-the-us-energy-system.
30. "Pathway to Critical and Formidable Goal of Net-Zero Emissions by 2050 Is Narrow but Brings Huge Benefits, According to IEA Special Report," IEA, May 18, 2021, https://www.iea.org/news/pathway-to-critical-and-formidable-goal-of-net-zero-emissions-by-2050-is-narrow-but-brings-huge-benefits.
31. "Pathway to Critical and Formidable Goal," IEA.

26. THE FUTURE

1. Joseph A. Schumpeter, *Capitalism, Socialism and Democracy* (New York: Routledge, 1976); originally published 1942 by Harper & Bros., New York.
2. Eastman Kodak Co., Form 10-K (Annual Report), Annual Reports, accessed February 10, 2022, https://www.annualreports.com/HostedData/AnnualReportArchive/e/NASDAQ_KODK_1998.pdf.
3. "Market Capitalization of Largest Companies in S&P 500 Index as of November 19, 2021," Statista, accessed February 10, 2022, https://www.statista.com/statistics/1181188/sandp500-largest-companies-market-cap/.

26. THE FUTURE

4. "A Green Bubble? We Dissect the Investment Boom," *The Economist*, May 22, 2021, https://www.economist.com/finance-and-economics/2021/05/17/green-assets-are-on-a-wild-ride.
5. Matthew Wilburn King, "Climate Change: How Brain Biases Prevent Climate Action," BBC, March 7, 2019, https://www.bbc.com/future/article/20190304-human-evolution-means-we-can-tackle-climate-change.
6. Kate Marvel, Twitter, https://twitter.com/drkatemarvel/status/1424359432578797574.

INDEX

Page numbers in *italics* indicate figures or tables.

CPSIA information can be obtained
at www.ICGtesting.com
Printed in the USA
LVHW101703241022
731418LV00022B/663/J